Data Analytics and Statistics

Andrew McDougall

First Edition

Bassim Hamadeh, CEO and Publisher
Gem Rabanera, Senior Project Editor
Casey Hands, Senior Production Editor
Emely Villavicencio, Senior Graphic Designer
JoHannah McDonald, Licensing Coordinator
Natalie Piccotti, Director of Marketing
Kassie Graves, Senior Vice President, Editorial
Alia Bales, Director, Project Editorial and Production

ISBN: 9798823338196

WEB-BASED RESOURCES

This book contains supplementary material available to complement your reading.

Please access the materials by clicking on this link:

https://cognella.app.box.com/s/n61he80z3z1g6oe4dhqtusv80jhdo355

Or access with electronic device by scanning QR code:

Contents

Preface

"There are lies, damn lies, and statistics[1] *"*

Welcome to statistics! A field of study where we can predict the winning candidate in an election before all the votes have been counted. We can also show that chocolate consumption is positively associated with traffic accidents. Both examples are supported by statistical evidence, but the chocolate consumption versus traffic accidents is spurious statistics.

The motivation for this book arose from teaching undergraduate and graduate students who required a methods-based course in Statistics for their major degree program. For most students, this would be their first real exposure to data analytics and statistical inference.

The biggest challenge for students is understanding **inferential reasoning.**

That is, Statistical methods can be used to make a **conclusion** about the **"population"** based on a statistical analysis of a (random) **sample** selected from the population. Since the sample only represents a subset of the population, we can never be absolutely certain our conclusion is true — students used to deductive reasoning[2] often find this disconcerting.

For students majoring in Statistics and Data Analytics, there is a wealth of appropriate textbooks that are available. For undergraduate students across a broad spectrum of disciplines who require a basic knowledge of Statistics for their major, this book is for you.

So what lies ahead?

Let's find out.

---

[1] The earliest printed version of this statement was in a Letter to the Editor of the National Observer on June 8, 1891 which was published June 13, 1891: *"Sir, It has been wittily remarked that there are three kinds of falsehood: the first is a 'fib,' the second is a downright lie, and the third and most aggravated is statistics..."* [Wikipedia]

[2] A deductive reasoning example is: Let A, B, C be three numbers "if $A = B$ and $B = C$ then $A = C$"

Our objective is to provide a textbook that caters to the actual material and knowledge that students need to achieve in an *Introductory Data Analytics and Statistics* course. That is, no "bells-and-whistles" *catchall* approach to topics that are never covered. Similarly, we try to minimize the "formula-based" presentation of Statistical methods by using software.

Software: In this book we use JMP (pronounced as "Jump") output to illustrate the results of a Statistical analysis. Alternative Statistical software certainly exist (see Appendix A), but the advantage of JMP is that it is menu-driven (no programming code required), and produces excellent graphics. SPSS is another popular Statistical software package and both JMP and SPSS can read Excel files and can be installed on PC and OSX (Mac) computers.

The key point is that at the level of this textbook, all the JMP results can be reproduced using other Statistical software packages.

Datasets: JMP and Excel versions of the datasets used in this book are provided. The **DATASET** name is given in the Exercise. The JMP or Excel file is denoted by chapter-name For example, ch3-pulse.jmp for the **PULSE** dataset in Exercise 3.2

TO BE DETERMINED — Cognella to provide link to a data repository.

What we cover in this book.

Chapter 1 Introduction and three (3) examples of scientific studies. How to correct your dataset and get it into JMP or SPSS.
Chapter 2 Data type. Why is this important?
Chapter 3 Looking at your data. To paraphrase (Orwell, G, 1945) "*all graphs are equal, but some graphs are more equal* (better) *than others.*"
Chapter 4 Probability. When should we use these methods?
Chapter 5 Correlation and Simple Linear Regression. When should we use these methods?
Chapter 6 Categorical analysis. The Goodness-of-Fit test and Contingency Tables
Chapter 7 Sampling distributions. Confidence Intervals and Hypothesis Tests basics
Chapter 8 Inference. Confidence Intervals and Hypothesis Tests for $t-$procedures
Chapter 9 Data analytics. Body roundness index versus BMI. Multiple linear regression.
Chapter 10 ANOVA. Analyzing more than two (2) independent groups

Appendix A Software notes
Appendix B Statistical Tables

Acknowledgments

Special thanks go to our project editor and to the Cognella production team for all their efforts in making this book a reality. I would also like to thank my wife Karen for her enduring patience, support, and timely reminders to "Finish it!" Needless to say, we have.

Montclair, New Jersey

Andrew McDougall
April 2024

Part I

Exploratory Data Analysis

1
Introduction

Let us begin our study of Statistics. Context is the critical issue. One does not conduct a study, survey, or experiment without having some *objective* or purpose in mind. That may seem fait accompli, so why do we then "throw" away the individual data we collect!?

Key Point: Statistics is about **summarizing** information

Of course, we don't actually throw away the data since we need it to calculate our *summary statistics*. The key point is that we want to make *conclusions* about the "study" based on a *statistical analysis* of the data and NOT on the individual values in the *sample* dataset.

The *italicized* words suggest Statistics has its own unique lexicon and implies we will need to develop skills in two important areas:

1. **Computation:** using software to create summary statistics and applying appropriate statistical analysis methods.

2. **Inference:** making conclusions based on our statistical analysis and writing reports that can be understood by our client.

1.1 Case Study: Survey

Example 1.1 Students attending Erehwon University were asked to complete a short survey requesting their `Gender, AGE, Gpa` and `CC debt` (credit card debt). A total of 100 students responded to the survey.

The aim of the survey is to see if any of the *"factors"* `Gender, AGE, Gpa` have an impact on `CC debt` for students attending Erehwon University. For example, is there a difference between Females and Males with regard to the amount of `CC debt`? Table 1.1 shows an excerpt of the results that were obtained.

TABLE 1.1. Case Study: Survey excerpt (" ⋮ " denotes rows 009 to 099 not shown)

Respondent	ID	Gender	AGE	Gpa	CC debt
aristotle	001	M		3.720	\$ 2M
bach	002	male	26	4000	0
NA	003	⋆	22	0.7	\$ 0.13
dino	004	f	52	4.0	
einstein	005	m	.	2.1	$e = mc^2$
flipper	006	Female	26	3.1416	\$ 35,000
garfield	007	F	18.5	0	–
hobbes	008	m	30	1.999	1.000,23
⋮	⋮	⋮	⋮	⋮	⋮
z0rro	100	m	−999	3.1	blank

Table 1.1 is a complete mess !!! However, it serves to illustrate some important components of data that we will discuss in more detail in chapters 2 and 3.

Typically, we would enter the **Raw** data shown in Table 1.1 into an Excel spreadsheet and then import the Excel spreadsheet into our desired **statistical software** package, such as JMP (pronounced as "Jump"). That is,

Raw data → **Excel** spreadsheet → **JMP** dataset

Remark: Some advantages of using Excel and JMP are:

1. An empty Excel spreadsheet is automatically created with rows and columns making it easier to enter the Raw data values in the correct place.
2. Excel exists for both PC and Mac[1] and spreadsheets are interchangeable (a PC Excel file can be read by Mac Excel and vice versa). Thus, the spreadsheet is *transportable.*
3. JMP also exists for PC and Mac and can import an Excel spreadsheet

Key Point: Software can read your data, but it cannot **"see"** your data.

For example, both Excel and JMP can read the character string $e = mc^2$ in Table 1.1 although **we** can immediately "see" this data entry for CC debt is not appropriate[2].

[1]PC refers to a computer that uses the Windows OS (Operating System). Mac refers to a computer that uses OSX

[2]Einstein's famous theorem, but not a valid data entry for CC debt

Excel Spreadsheet

Figure 1.1 shows the Excel spreadsheet `mydata.xlsx` that was transcribed from the **Raw** data in Table 1.1 (ignoring the omitted rows).

> *Clearly, we made some substantial changes in transcribing the Raw data entries to the Excel spreadsheet.* — Why?

	A	B	C	D	E	F
1	Respondent	ID	Gender	AGE	Gpa	CC debt
2	aristotle	001	m		3.72	2000000
3	bach	002	m	26		0
4		003		22	0.7	0.13
5	dino	004	f	52	4	
6	einstein	005	m		2.1	
7	flipper	006	f	26		35000
8	garfield	007	f	18.5	0	
9	hobbes	008	m	30		
10	z0rro	100	m		3.1	

FIGURE 1.1. Excel spreadsheet from Table 1.1

The simple answer is that the data entries shown in Table 1.1 need to be **"cleaned"** in order to provide useful results for analysis purposes. For example, what is the proportion (percentage) of males versus females who responded to the survey? Without re-formatting Table 1.1 this would be a very tedious task. So what do we need to do? Specifically:

- **Missing Values** handled consistently and/or inserted for obvious errors
- **Obvious Errors** corrected (if possible) — otherwise set to missing
- **Format Consistency** such as **only** "m or f" (or missing) used for Gender

Our objective is to import the Excel spreadsheet `mydata.xlsx` into JMP which depends on knowing the **data type** of the data entries in each **variable** (column). We discuss data types in more detail in chapter 2, but for now we first note some general attributes of the Excel spreadsheet shown in Figure 1.1 (Mac version).

1. The first row of the Excel spreadsheet `mydata.xlsx` is often used for the column headers Respondent, ID, Gender, AGE, Gpa, CC debt since JMP can use these as column headers in the JMP dataset `mydata.jmp` [see Figure 1.2]

2. Excel recognizes **categorical** variables such as Respondent and Gender and makes all the data entries left-flushed in each cell. In contrast, Excel makes the data entries of the **numeric** variables AGE, Gpa, CC debt right-flushed.

3. When we use *numeric* data entries as "labels," such as the ID variable, Excel needs to be coerced to designate these entries as categorical.

Missing Values: Surveys will often contain *"missing values"* where the participant does not provide a response such as no AGE for aristotle, ID = 001. However, Table 1.1 uses a variety of entries for missing values:

NA, $\star$, [empty space], . (period), −999, −, and the word "blank"

Key Point: The best approach is to use an [*empty space*] in the Excel spreadsheet cell.

Obvious Errors: AGE = −999 is obviously impossible, but it was used as a code for a "missing value" (not recommended since all the "−999" entries will need to be recoded). Other errors may be more difficult to detect since we need to use *contextual* or external knowledge.

- Gpa should be a value between 0 and 4.0 so bach's Gpa=4000 is an obvious error.

- Gpa is *rounded* to three (3) decimal places so flipper's Gpa=3.1416 is not correct.

- What about garfield's Gpa=0 ? [Okay if this is garfield's first semester.]

- On the other hand, hobbes' Gpa=1.999 is an error since it can't (realistically) be achieved — see Exercise 1.1

- Lastly, hobbes' CC debt = 1.000, 23 just looks like a typo (the "." and "," are reversed). However, some European countries use this monetary notation so this could be valid.

Question: *What should we do with obvious errors (or potential typos) ?*

Go back to the Raw data and check the Excel file `mydata.xlsx` for typos or incorrect entries (e.g., suppose we entered "185" for garfield's AGE instead of 18.5). Correcting our typos is easy — but time-consuming. So be diligent with your transcription of the Raw data.

Key Point: Unresolved errors should be set to **missing**

Format Consistency: Excel can accept essentially any data entry whereas JMP is more strict about the data entry **format.** For example, JMP would consider the Gender entries shown in Table 1.1 to have seven (7) different "levels" (labels):

M, male, $\star$*, f, m, Female, F*

Clearly not what we want. Similarly, JMP would treat **all** the CC debt entries as *character* due to the $ sign, the word blank, hyphen –, etc. Hence these were changed to numeric in Excel by deleting these character entries.

Key Point: Use a **consistent** format. No characters mixed into numeric data

JMP Dataset

Importing a cleaned Excel spreadsheet into JMP is relatively straightforward:

File → Open Select `mydata.xlsx` then **Import** in the *dialog* popup window

For the Excel spreadsheet in Figure 1.1 the resulting JMP dataset is shown in Figure 1.2

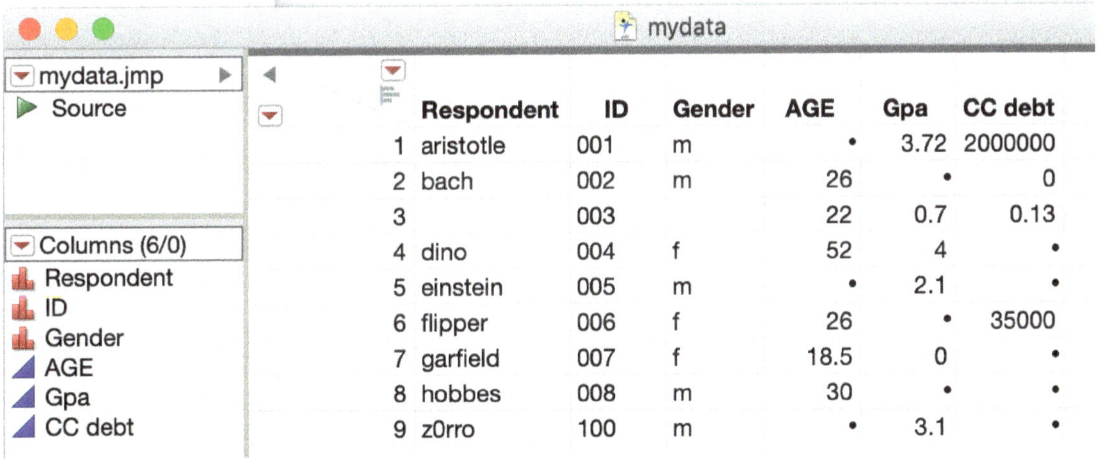

mydata

mydata.jmp
Source

Columns (6/0)
Respondent
ID
Gender
AGE
Gpa
CC debt

	Respondent	ID	Gender	AGE	Gpa	CC debt
1	aristotle	001	m	•	3.72	2000000
2	bach	002	m	26	•	0
3		003		22	0.7	0.13
4	dino	004	f	52	4	•
5	einstein	005	m	•	2.1	•
6	flipper	006	f	26	•	35000
7	garfield	007	f	18.5	0	•
8	hobbes	008	m	30	•	•
9	z0rro	100	m	•	3.1	•

FIGURE 1.2. Excel file imported to JMP Dataset for Table 1.1

The JMP dataset has some minor differences from the Excel spreadsheet.

- The column names (variables) were imported from the first row of the Excel spreadsheet
- Missing values for numeric variables are denoted by • entries. [See Appendix A]

Question: *Why not just enter the raw data into JMP ?*

We could, but JMP is proprietary software so if we sent our `mydata.jmp` file to our client for review, they would need to have JMP installed on their computer in order to open this file.

Key Point: JMP datasets can be converted to an Excel (or Text) file.

SPSS Dataset

Another popular statistical software package is SPSS (Statistical Package for the Social Sciences) which also exists for PC and Mac. Like JMP, it is straightforward to import a cleaned Excel file into SPSS:

Open another file	Select `mydata.xlsx` in the *dialog* window, then click **OK**

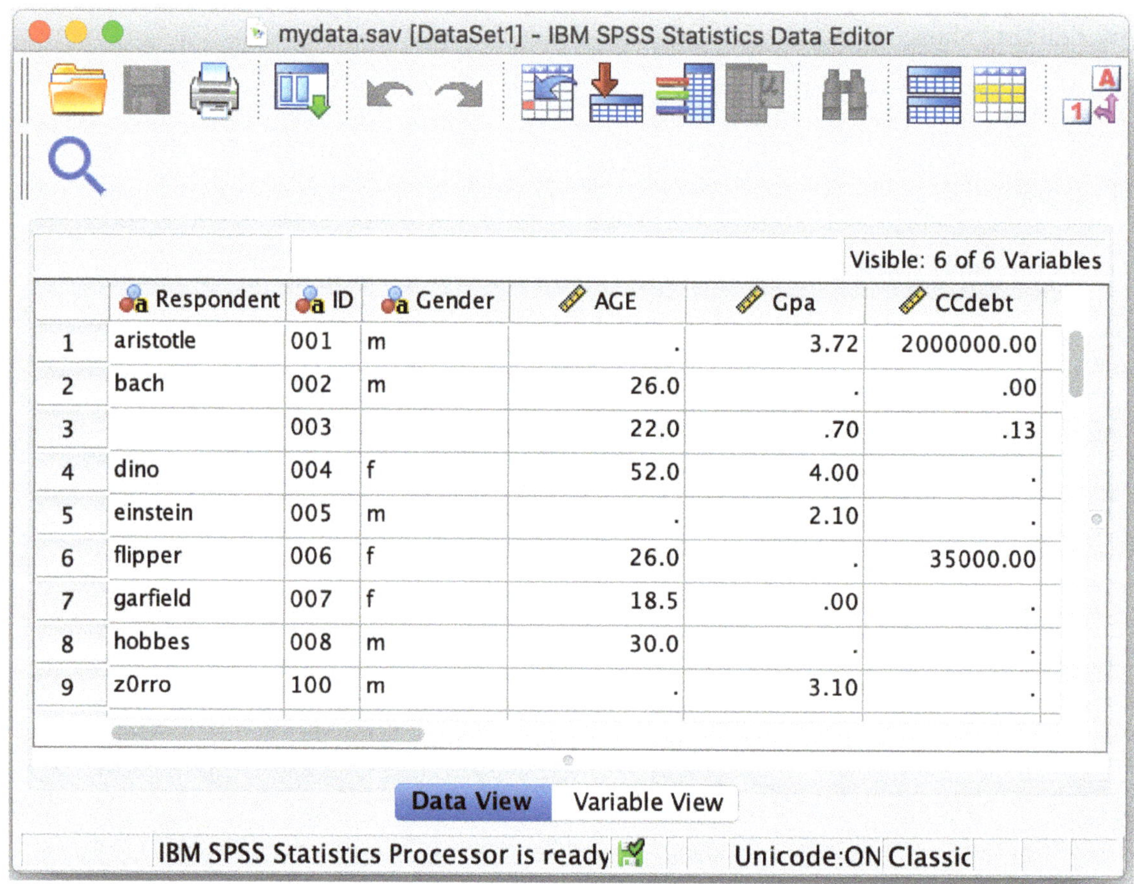

	Respondent	ID	Gender	AGE	Gpa	CCdebt
1	aristotle	001	m	.	3.72	2000000.00
2	bach	002	m	26.0	.	.00
3		003		22.0	.70	.13
4	dino	004	f	52.0	4.00	.
5	einstein	005	m	.	2.10	.
6	flipper	006	f	26.0	.	35000.00
7	garfield	007	f	18.5	.00	.
8	hobbes	008	m	30.0	.	.
9	z0rro	100	m	.	3.10	.

FIGURE 1.3. Excel file imported to SPSS Dataset for Table 1.1

The result is shown in Figure 1.3 and is similar to the JMP dataset — periods "." are used for missing numeric entries and an *empty cell* for missing categorical (character) entries.

A minor difference is that numeric variables (such as CCdebt) use the *same decimal format* for **all** entries. This does not impact the statistical analysis of the data.

Remark: Other statistical software packages are available which are listed in Appendix A. However, both JMP and SPSS are user-friendly and menu driven (meaning you can select the analysis procedure from a drop-down menu and then select options from a popup *dialog* window). We will use JMP output throughout this book.

1.2 Case Study: Experiment

The Survey case study in Example 1.1 is an example of an **Observational study** where we collected the responses from 100 students (the **sample**) out of the total number of students who were currently attending Erehwon University (the **population**). While we focused on how to get the observed "sample" data into a suitable format *after* it was collected, the question is

> *"how did we* **choose** *our participants?"*

This is really asking us to **PLAN** our collection procedure *before* we conduct the actual survey. This is referred to as the **study design** and can require considerable effort on our part since we need to anticipate potential issues *before* we conduct the study.

Key Point: A poorly designed study can result in **gigo** *garbage in = garbage out*

The alternative to an observational study is called an **Experimental study** where we **"actively"** attempt to induce a change in the participant's response. Clinical trials are examples of experimental studies where some patients are provided with a new drug or vaccine (the **treatment**) and other patients receive a placebo[3] as in the following example.

Example 1.2 Patients with elevated cholesterol levels are randomly assigned to two groups. Group A receives a new cholesterol drug that claims to lower LDL (bad cholesterol). Group B receives the "placebo" drug. Baseline cholesterol at the start of the study is compared to the patients' cholesterol level after three months. There were 100 patients in each group.

The *sample* consists of the *before-and-after* cholesterol measurements from the 200 patients in the study. Since the purpose of the study was to see if the new drug was effective in lowering LDL, the **target** *population* would be *all* subjects with elevated cholesterol levels

This example illustrates the following four basic components of a statistical study:

- **STATE** What is the purpose of the study (objective)
- **PLAN** How the objective is to be assessed (design)
- **SOLVE** Use appropriate statistical methods to analyze the data (results)
- **CONCLUDE** Use the results to make conclusions (inference)

We are not ready yet to deal with all these components of a statistical study. However, Example 1.2 illustrates that **context** is critical. That is, we need to understand some of the **science** involved in the study (e.g., what is considered an "effective" reduction in LDL?)

[3] A placebo is a "dummy" pill which looks like the real drug, but without the active ingredient. (Also euphemistically referred to as a "sugar pill.")

1.3 COVID-19: Data Analytics

The COVID-19 virus has dramatically impacted our lives. Social distancing, quarantines, masks, lockdowns, testing, remote learning, and re-opening are all terms that we have never experienced on such a world-wide scale. The rapid spread of COVID-19 and increasing number of deaths required national and state action. Figure 1.4 shows the timeline of the number of total daily deaths reported to the New Jersey DOH (Department of Health) during the initial surge in early 2020. So let us breakdown this *graphical* summary with **data analytics** which uses statistics to make decisions.

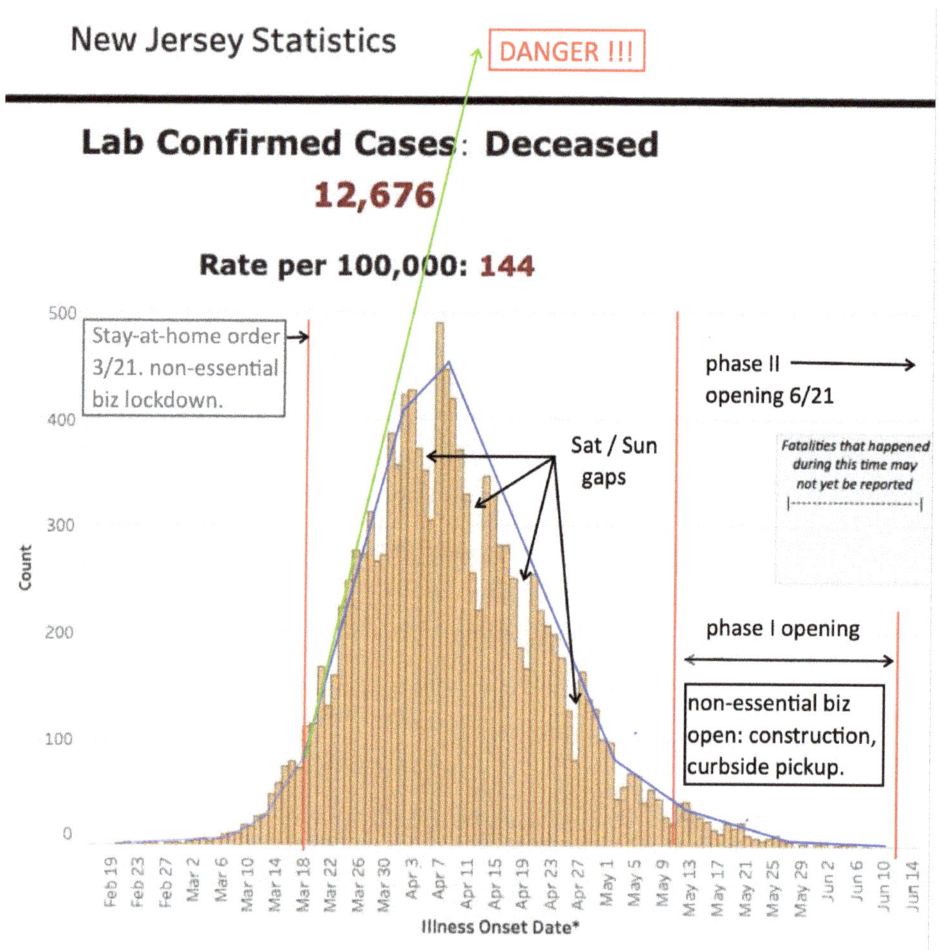

FIGURE 1.4. COVID-19 Deaths in NJ: February to June 2020

We begin with the mantra quoted by Federal and State Health agencies:

> *"Look at the data. Follow the science."*

The problem is that COVID-19 was a new virus and so very little *science* was initially known — but there was **data**! It became increasingly clear that COVID-19 was a global pandemic and causing a rapid rise in deaths. Action was obviously required even if the science was limited. Hence the imposition of Executive Orders (EO) issued by NJ State as indicated by the red vertical lines and annotated text boxes shown in Figure 1.4.

> What did the NJ data show us in *real-time*.

Visualization The graph (orange vertical "bars") shows the number of daily confirmed COVID-19 deaths in NJ. The height of the bar is calibrated by the left axis `Count` so we see that the peak daily number of deaths was almost 500 before the decline. This is an example of a **barplot** (also called a bar graph and bar chart). Now we turn our attention to the annotation.

Decision What if NJ did nothing? The green line is a *projection* (prediction) of the number of future deaths based on the data leading up to the stay-at-home order[4] EO 107 issued on March 21, 2020 (first red vertical line on the left side of Figure 1.4). It was not the only data analytic projection, but tracking COVID-19 deaths made it very clear that doing nothing would be **unacceptable**.

Conclusion Prior to the EO 107, NJ had already started to aggressively implement "social distancing" (no gatherings) and switch to remote learning for all K-12 Schools, Colleges and Universities. The stay-at-home order also closed all nonessential retail businesses (despite the economic impact). That is, decisions were made to introduce *interventions* to try and curb the spread of COVID-19 with the primary aim of reducing the number of hospitalizations and deaths that were occurring at an alarming rate.

Question *Did it work?*

Yes, but it took time as shown by the blue curve in Figure 1.4. Reporting deaths on weekends appeared to be delayed (dips indicated on the graph). The blue curve did "bend" and partial re-opening of businesses (Phase I and II) occurred in NJ and other states.

Key Point: Statistics and Data Analytics use **data visualization** extensively

A picture is worth a thousand words[5] A simple graph — a powerful tool. We will learn how to produce effective data visualization graphs and, more importantly, learn how to interpret the statistical features of these graphs relevant to the context of the study.

---

[4]How did we get this projection? Cover up the right side of the graph from the 3/21 stay-at-home order. Given the "trend" up to this point, extending a simple straight line was clearly heading towards the DANGER !!! zone.

[5]See `https://en.wikipedia.org/wiki/A_picture_is_worth_a_thousand_words` for historical variations of this phrase

Chapter 1 Exercises

1.1 Refer to Table 1.1. Assume hobbes gets a C (Gpa=2.0) in all his (3 credit) courses except one in his first semester where he got a C− (Gpa=1.7). Thus, $(n-1)*2.0+1.7 = n*1.999$ where n = total # of courses.

a) How many courses would hobbes need to take to get a Gpa=1.999 ? [Solve for n]

b) If hobbes does 5 courses per semester (fall and spring only), how long will it take?

1.2 Supermarkets carry a wide variety of product lines and brands which need to be restocked on a regular basis. Using data analytics, a supermarket can assess which product line, brand, and layout is profitable. Do a web search to answer the following questions.

a) How do supermarkets make you spend more money?

b) How does the supermarket layout impact the customer psychology?

1.3 Refer to Example 1.2. Patients with elevated cholesterol may need different dosages of the new drug depending on their initial LDL level.

a) How would you allocate the 200 patients if three (3) dosages were available: placebo, low dose, high dose? [Remember, you can not have a "fraction" of a patient.]

b) Is there an issue with assigning patients who have elevated cholesterol levels to the placebo group? Explain. [HINT: *"Why do we need a placebo group?"*]

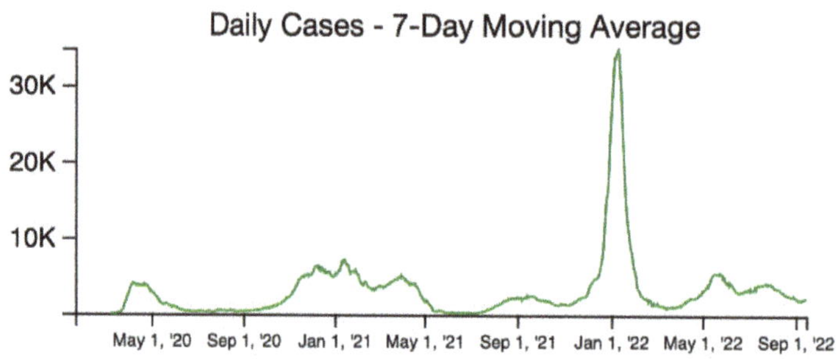

FIGURE 1.5. COVID Cases and Deaths in NJ: January 21, 2020 to September 15, 2022

1.4 Figure 1.5 shows daily COVID cases and deaths in NJ from the initial pandemic through to September 2022. The COVID-19 vaccines were widely available in the US by April 2021 and were found to be effective against the COVID mutations that evolved.

a) As can be seen, NJ experienced two additional surges in daily deaths. Noting the general timeframe of these surges, what are some possible reasons for this?

b) Using a ruler, argue that the dramatic increase of daily cases in early 2022 provided a **leading indicator** for daily COVID deaths.

1.5 A study of three (3) Antidepressant drugs is planned where 18 patients who suffer from depression are to be randomly assigned to one of the three drugs. The researcher proposed the following design based on Gender and Age groups as shown in Table 1.2.

a) Six (6) different patients were assigned to each Antidepressant drug. Why does this make sense from an information perspective.

b) Do you support the Researcher's proposed design? Explain why not.

TABLE 1.2. Antidepressant Data: Proposed Design

	Patient	Gender	Age group	AntiDep
1	1	F	<20	A
2	8	F	20-40	A
3	16	M	>40	A
4	15	F	>40	A
5	10	F	<20	A
6	3	F	<20	A
7	5	M	>40	B
8	2	F	20-40	B
9	4	F	>40	B
10	14	M	20-40	B
11	11	M	20-40	B
12	7	M	20-40	B
13	6	M	>40	C
14	18	M	<20	C
15	9	M	<20	C
16	17	F	>40	C
17	12	M	<20	C
18	13	F	20-40	C

2
Data

Data are measurements recorded from a *sample* collected from a *population.* The Survey case study and Cholesterol experiment presented in the previous chapter are two examples where only part of the **target** population was (somehow?) selected and *specific* data measurements were recorded from the sample. Both these studies can be adapted to the **4 Step Process** we introduced in section 1.2 and repeat here with emphasis on *when* we see the **DATA**

- **STATE** What is the purpose of the study (objective)
- **PLAN** How is the objective to be assessed using careful implementation (design)
- **SOLVE** Use appropriate (statistical) methods to analyze the **DATA** (results)
- **CONCLUDE** The sample is used to "infer" conclusions about the population

That is, the quality and validity of the **DATA** we collect clearly depend *a priori*[1] on how well we developed the study design (the **STATE** and **PLAN** steps). The point here is that collecting data is *"expensive"* — there can be real costs involved and it always takes "time" to record individual measurements. Data that is **gigo**[2] is obviously a waste of time and money and can lead to completely meaningless results (**SOLVE** step) and misleading conclusions about the population.

This all seems rather complicated which raises the following question.

Question: *Why not just measure the* **entire** *population?*

[1]Latin for — *from what is before* — here, we are saying that the **study design** comes **before** we collect our **DATA**
[2]See section 1.2 *garbage in = garbage out*

2.1 Population versus Sample

There are some situations where we can measure the *entire* **population** where data is collected from *every* subject or item in the population. Here are some examples.

Lottery: The total number of combinations in a Lottery game is known, so the chances of winning can be computed. Similarly, the odds of getting any particular type of Poker hand is known (as with other casino gambling games). These are applications of **probability** where the number of possible "outcomes" (the population) is already known without the need to collect data. *This situation is not the topic of interest (data) in this chapter.*

Census: An important example of a measured population is the US decennial (10-year) census which collects data from *every* person currently living[3] in the US (including US territories). This is VERY expensive (estimated to cost $15.6 billion for the 2020 census) and the schedule[4] for releasing the census State population counts (as well as other results) was delayed due to the impact of COVID-19.

Impractical: On the other hand, there are situations where measuring an entire population is impractical (even if it "could" be done). For example, car manufacturers are required to perform safety crash tests for the cars they produce. Clearly, crash testing "every" car the manufacturer produces (a known *finite* number) would be ridiculous!

Impossible: Carp is an invasive fish species that is impacting the survival of other fish species in the Great Lakes and Mississippi river[5] (opinion article, but worth reading). Carp live underwater, so we would have to agree that counting the total number (current population of Carp) would represent an impossible situation. Fortunately, *catch-and-release* **sample** strategies enable us to monitor the carp problem (and thus make remediation decisions).

Key Point: Data is recorded from a (statistical) **sample** collected from a **population**

The term "statistical" refers to the **study design** where we need to decide *how many* subjects to measure, *what* characteristics (variables) to measure, and *who* gets selected. So there are some disadvantages — but also many advantages — in using a sample.

Sample Disadvantages

- The sample may not properly represent the intended (target) population. A **biased** sample (e.g., you only sampled students in your classes at Erehwon University).
- Some important variables are not well-defined or were not measured. Data **quality** (e.g., how is `CC Debt` defined?) If `Height,Weight` were not measured in the Cholesterol study — then no BMI (Body Mass Index) available for assessing obesity.

---

[3]Citizens, foreign residents, or otherwise. Tourists and foreign business travelers are not counted in the US census.

[4]https://www.census.gov/programs-surveys/decennial-census/decade/2020/planning-management/operational-adjustments.html

[5] https://www.crainsdetroit.com/environment/michigan-beyond-11th-hour-asian-carp-move-closer-great-lakes

Sample Advantages

- Samples are cost and time effective. Results are available in a short time-frame.
- Sample size (n) does not depend on population size (for N large). This may seem surprising, but a poll of $n = 1500$ respondents will suffice for any N (large).
- Sample data is often *more* accurate than population data. It is easier to transcribe sample data and correct errors (or set to missing). See Figure 1.1

2.2 Datasets

The **SOLVE** step shows that we need to have collected the **raw** data for each subject in the study sample. This is typically collected and recorded in two ways:

- From hand-written records or existing databases
- From software applications that automatically create an electronic file record

In either case, individual subject data will need to be entered into a **master** database such as a plain text file (TextEdit or NotePad) or an Excel spreadsheet (do NOT use Word). These formats allow the master database to be imported into a statistical software package such as JMP to create a **dataset**. To summarize the features of Figures 1.1 and 1.2

- Excel (Mac or PC). Has some limited statistical analysis methods
- JMP (Mac or PC). Menu-driven. Can import Plain text and Excel files

2.3 Data Types

Table 2.1 shows a JMP dataset from an Antidepressant study which includes several variables that have different **data types**. For example, **Gender** has *character* entries and **PostTrt** has *numeric* entries. Thus, we define a variable to have a data type which is either:

Categorical	Character entries. Left flush in JMP [**Gender, Age, AntiDep**]
Quantitative	Numeric entries. Right flush in JMP [**Block, PostTrt, PreTrt**]

The distinction between a categorical variable and a quantitative variable is important in JMP since it determines the type of results generated in the **SOLVE** step. JMP will assume a variable is categorical if any data entry contains a character. Otherwise it is quantitative.

Example 2.1 What if `Gender` had been coded as 1 (Female) and 2 (Male) in Table 2.1 ?

TABLE 2.1. Designed Experiment: Antidepressant Data

	Block	Gender	Age (years	AntiDep	PostTrt	PreTrt
1	1	F	<20	A	21	48
2	2	F	20-40	A	22	43
3	3	F	>40	A	18	44
4	4	M	<20	A	26	42
5	5	M	20-40	A	21	37
6	6	M	>40	A	18	41
7	1	F	<20	B	25	36
8	2	F	20-40	B	21	31
9	3	F	>40	B	24	35
10	4	M	<20	B	20	38
11	5	M	20-40	B	24	34
12	6	M	>40	B	24	36
13	1	F	<20	C	17	31
14	2	F	20-40	C	19	28
15	3	F	>40	C	18	29
16	4	M	<20	C	17	29
17	5	M	20-40	C	15	28
18	6	M	>40	C	19	26

JMP would then consider `Gender` to be a *quantitative* variable and produce results that are not appropriate for analyzing this data type[6] So we examine the distinction between a categorical variable and a quantitative variable in more detail.

Quantitative Let x and y be two numeric data values. For example, `PostTrt` = 15, 19 (last two entries in Table 2.1) A quantitative variable must satisfy the following:

1. $x < y$ is meaningful — `PostTrt` $= 15 < 19$ [Patient # 17 did better than #18]
2. $x - y$ is meaningful — $(15 - 19) = -4$ [# 17 was 4 points lower than #18]

[6] The "data type" can be easily changed in JMP. Numeric encoding is often used for variables such as Gender in legal discrimination cases to avoid bias on the part of an outside expert witness (the Statistician) who analyzes the data without knowledge of which Gender is represented by 1 or 2

Categorical The data entries for a categorical variable are simply **labels** so "M" < "F" or "M" − "F" makes no sense. Even if we encoded F, M as $1, 2$ what does $2 - 1 = 1$ mean ? The difference between a Male and Female = Female !?

2.4 Ordinal Data

The `Age` variable in Table 2.1 is categorical, but has an "inherent" **ordering** associated with the three age groups. That is, we would agree that "<20" *is less than* "20–40" in terms of `Age` group, but that the difference "<20" *minus* "20–40" is not meaningful. So `Age` is partly quantitative, but still categorical (we could label the `Age` groups in any way we like).

Ordinal Data Type Categorical variable that has an *inherent* ordering of groups

Likert Scale This is a psychometric (ordinal) scale that is commonly used in questionnaires where respondents answer a series of items based on a *rating scale*. For example, a 5-point Likert scale typically has the (symmetric) format shown in Table 2.2.

TABLE 2.2. Example of a 5-point Likert Scale format

	Please check your response	Strongly Agree	Agree	Neutral	Disagree	Strongly Disagree
Q1.	JMP was easy to install	○	○	○	○	○
	⋮					
Qx.	*More questions*	○	○	○	○	○

Similarly, 7-point and 10-point scales (Never Recommend — Highly Recommend) are used for feedback surveys. The main issue is that these ratings are **not** necessarily *equidistant* even though they are often analyzed numerically as 1, 2, 3, 4, 5 That is, it assumes **all** respondents "interpret" the choice between Strongly Agree versus Agree in *exactly* the same way as the choice between Agree versus Neutral, Neutral versus Disagree, and Disagree versus Strongly Disagree. Research[7] has shown that respondents tend to be less likely to choose the *extremes* Strongly Agree, Strongly Disagree versus the *perceived* more moderate ratings.

2.5 Missing Values

In practice, studies may have some *missing* values. We addressed the data *format* for assigning a data entry as missing with respect to Table 1.1 and Figure 1.2 in the `CC Debt` Survey case study. In general, missing values do not present a problem if they constitute less than 5% of the dataset. Some examples where missing data can arise are:

[7] For further details, see https://en.wikipedia.org/wiki/Likert_scale

- A subject does not respond to certain survey questions (like `CC Debt`)
- Some trials in an experiment fail to produce a result (e.g., Suppose subject #16 in Table 2.1 did not return to the study. Then **PostTrt** would be missing)
- Some trials were simply not performed (e.g., no suitable subject was found for row=16)
- There is an obvious error in the recorded dataset such as pH = 0.8 (hyper-acidic) when measuring sea water at a particular location (normal seawater pH is close to 7).

JMP (and other software) will ignore any missing data value entry. That is, JMP will still perform the statistical analysis we request (adjusted for missing data values). This is usually not a problem when the (overall) proportion of missing data values is less than 5% However, the following example illustrates how missing data values can impact our statistical analysis

Example 2.2 Missing Values A survey has 20 questions. Suppose 5% of 1000 responses are "randomly" missing for each question. Then the chance that *none of the 20 questions has a missing value* = $(1 - 0.05)^{20} = 0.358$ That is, we can only expect to get 358 "complete" responses out of 1000 surveys.

2.6 Auxiliary Variables and Covariates

In the Cholesterol experiment (**Sample Disadvantages**) we indicated that a patient's BMI could be obtained "if" we had recorded their `Height` and `Weight`. Note that the formula for BMI depends on the **units** used to measure height and weight as follows:

$$\text{BMI} = \frac{\text{Weight(kg)}}{[\text{Height(meters)}]^2} = 703 * \frac{\text{Weight(lbs)}}{[\text{Height(in)}]^2}$$

Since people know their height (H) and (true) weight (W), but may not know their BMI, we can use JMP to create the **auxiliary** variable BMI from the recorded H and W data values and add the `BMI` "column" to our existing dataset. That is, we provide JMP with the BMI formula above and simply get JMP to "Apply" this to get the column of BMI data values.

In similar fashion, we can **categorize** a numeric variable such as person's actual `Age` into groups as shown in Table 2.1 or use the "new" BMI variable to create a classification variable, `cBMI` say, with character entries `underweight, normal, overweight, obese` We will explore the "formula" functionality of JMP in subsequent chapters.

Covariates are variables that have been been recorded in the dataset and are considered useful for analyzing the objective of the study. For example, in the Antidepressant dataset (Table 2.1) **PreTrt** provides a "baseline" measure of the depression score for each subject prior to administering the **AntiDep** drug. Our interest is whether there is a statistically significant difference (or not) in the **PostTrt** depression scores in patients from their **PreTrt** scores. Similarly, **Gender** and **Age** are covariates that may show whether there are statistically significant differences due to these variables with respect to **PostTrt** - **PreTrt** On the other hand, we note that **Block** is an Auxiliary variable since it was derived from the six (6) **Age** and **Gender** combinations. In summary, for the Antidepresant data in Table 2.1

Response:	Variable of primary interest [**PostTrt**]
Covariate:	Recorded variables of interest [**Gender, Age, PreTrt**]
Auxiliary:	Variable derived from Covariates [**Block**]

Key Point: Only relevant Covariates should be selected

For example, "Eye Color" could have been included in the Antidepressant study, but we would agree with the researcher that this covariate is unlikely to be relevant to the depression score of a subject. We can help with the design, but ultimately we rely on the researcher to define the variables of interest to the study.

Chapter 2 Exercises

2.1 A Social Security number (SSN) is a unique nine-digit number, `AAA GG XXXX`, issued to US citizens, permanent residents, and temporary (working) residents.
(See http://www.ssa.gov for details.)

a) Prior to June 25, 2011 the "Area" code `AAA` of the SSN was assigned *lower* numbers to applications received from east coast states and progressed to *higher* numbers for the west coast states. What data type do SSNs represent here?

b) From June 25, 2011 onwards the "Area" code `AAA` identifier was discontinued and SSNs were randomized (with certain exclusions such as 000, 666, and the 900+ series). Does this change the data type of SSNs? Explain.

2.2 Dummy variables are binary (0,1) auxiliary variables that can be used to encode categorical variables with $k > 2$ groups[8] such as marital status MS ($k = 4$) = single, married, widowed, divorced.

Show that three ($k-1$=3) dummy variables, $M1, M2, M3$ say, can uniquely define the $k = 4$ MS classes. [HINT: At MOST **one** of $M1, M2, M3$ can be coded as "1"]

2.3 Suppose a survey question has k options where you can *"Select all that apply"*

a) Show that each option would need to be encoded separately by a dummy variable.

b) Discuss the difference between the options *"None of the above"* versus *"Decline to answer"* [HINT: Why does the latter choice **not** provide "usable" information]

[8] Gender (k=2) is a dummy variable, but usually encoded as m, f for clarity

2.4 Fake data

The height (in inches) of 100 college students were measured. The first 10 results are:

61.70160 61.40626 77.56282 62.00116 74.51498

70.35044 59.96617 70.90227 73.03435 62.40639

Why do you know this is fake data? Explain. [HINT: Width of a strand of hair?]

2.5 Simulated data is used extensively for assessing predictive models, conjectures, and theoretical results. A simple case is where we want to generate a set of **random numbers** Describe how you would generate a random sequence of 20 integers, each taking a value between 1 and 6 inclusive. For example 41163... [HINT: Die toss]

2.6 Before-and-After experiments produce data such as Table 2.1 where subjects are measured *before* receiving the "treatment" and then *after* the treatment was administered. Consider the following weight loss data for two (2) different diet regimes

a) Which diet produced the greatest overall weight loss?

b) Show the *percentage* weight loss for **each** subject was 20%

c) Hence, what is your conclusion about the two diets?

TABLE 2.3. Weight Loss Data

Patient	Diet	Before Wgt	After Wgt
1	A	250	200
2	A	150	120
3	A	125	100
4	B	300	240
5	B	100	80
6	B	200	160

2.7 A marine biologist was interested in assessing the effect of a "predator" (lobster) on the size and number offspring produced by female marine snails under controlled conditions. That is, the snails were protected from the lobster (if present) in each aquarium and the temperature, salinity, oxygen, and food supply were maintained at constant levels during the experiment. Prezant et al. (2006)

a) What **variables** needed to be recorded for the researcher to assess the **objective**?
[HINT: How should we measure "size" of the (3-dimensional) offspring.]

b) The sex of a marine snail is difficult to determine (noninvasively) and so some male snails were included in the experiment.

What was the consequence of this?

3

Looking at Data

The variables in a dataset have two basic data types: categorical and quantitative. In this chapter we consider some standard statistical techniques for univariate EDA (Exploratory Data Analysis). That is, we "look" at each variable in the dataset separately to assess the statistical features of interest. The categorical case is discussed in the next section. For a quantitative variable that we generically label as "X" three EDA components are involved.

- **SHAPE:** Graphical visualization of the distribution of X
- **LOCATION:** Numerical summary statistics describing where X is located
- **SPREAD:** Numerical summary statistics describing the variability in X

3.1 Categorical Variable

3.1.1 Barplot

For a categorical variable, such as `GENDER`, the graphical visualization technique we will employ is the Barplot. Synonymous terms are Bar Chart and Bar Graph. To create a barplot, we first count the total number of `Female` and `Male` subjects to get a Frequency Table. For example, the frequency table result for the "handwritten" dataset is given below.

```
M M F M F    F M M M F    F F M M M    M M F M F
```

GENDER	Tally	Frequency	Proportion
F = Female	~~////~~ ///	8	0.4
M = Male	~~////~~ ~~////~~ //	12	0.6
Total		20	1.0

From the Frequency Table the Barplot represents the `Female` and `Male` groups as "bars" of height = Frequency as shown in the JMP output of Figure 3.1

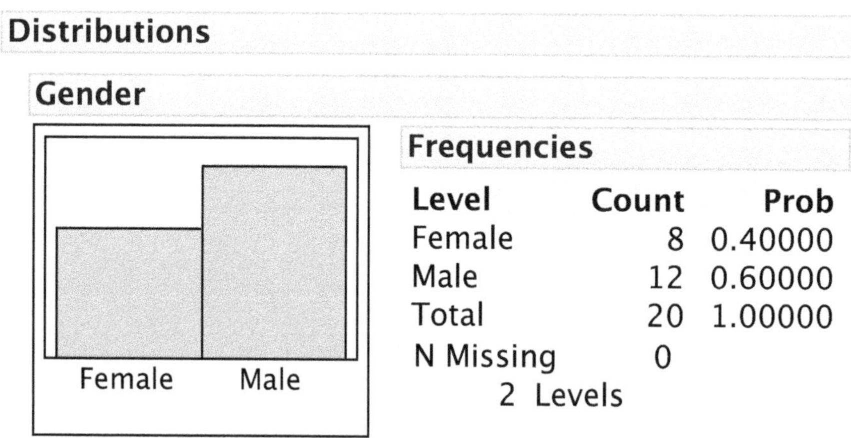

FIGURE 3.1. Gender Barplot

Key Point: Shape, Location, and Spread statistics are **not** relevant for categorical variables

This is because the "labels" `Female` and `Male` are essentially "arbitrary" in the sense that we could have put the `Male` "bar" first without changing our interpretation of Figure 3.1

Interpretation of a Barplot
We compare the relative *bar heights* across the different groups for a categorical variable.

This also applies to **ordinal** data, save that we would display the groups according to the inherent ordering.

3.1.2 Pie Charts

These are used extensively for colorful presentations, but they are statistically inefficient. The following example shown in Figure 3.2 illustrates why we will not use Pie Charts.

The Barplot is clearly more "efficient" (visually and therefore statistically) since even slight differences in the *vertical* heights of neighboring bars is easily discernible. Note that this *linear* comparison was visually enhanced by displaying the groups in descending order of bar heights. (JMP can also display the groups in ascending order.) Because a Pie Chart is based on *angular* comparisons (which are nonlinear), differences can be visually impossible to detect. Additional numerical information is required for the Pie Chart that is not needed (visually) for the Barplot.

Key Point: Use Barplots for EDA of categorical variables

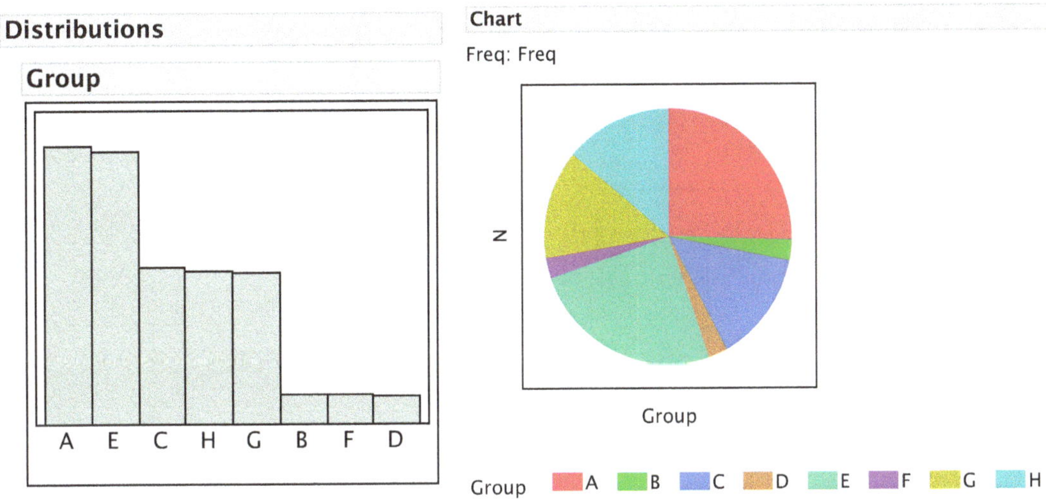

FIGURE 3.2. Barplot versus Pie Chart

3.2 Quantitative Variable

3.2.1 Stemplot

A Frequency Table may sometimes be useful for examining a quantitative variable, but it does not present the data values on a true calibrated scale. This is required to properly interpret the **SHAPE** of the distribution of X. A simple way to do this is called a **Stemplot** (also referred to as a *Stem-and-Leaf* plot) which can be thought of as a "calibrated" Frequency Table. This is illustrated by the following example.

X	Tally	Frequency	Proportion
3.1	//	2	0.2
3.4	//	2	0.2
3.7	/	1	0.1
4.1	///	3	0.3
5.3	/	1	0.1
7.2	/	1	0.1
Total		10	1.0

The Frequency Table starts at 3.1 and ends at 7.2 To calibrate this table, we "split" the individual data values into a *Stem* unit and trailing *Leaf* digit representation: 3.1→3| 1 But since we have two 3.1 data values, they are displayed on the *same* Stem: 3.1 , 3.1→3| 1 1 All the other data values of X are split in similar fashion. Although we have no 6.x data values, we **must** still include the blank Stem: 6| to obtain a true calibration. Stacking up the Stem and Leaf representations of the data values gives the following Stemplot below.

Note that we have used `Courier` font since this assigns equal width to each number and so all the Leaf digits are aligned properly.

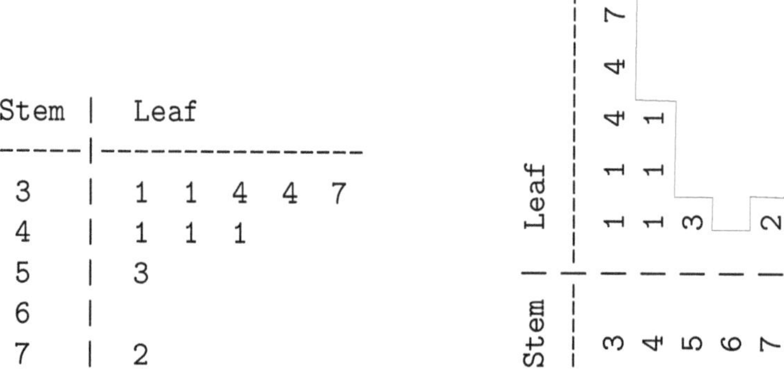

FIGURE 3.3. Stemplot and Profile rotated anti-clockwise 90°

Rotating the Stemplot anti-clockwise and using the Leaf digits *Profile* as the bar heights results in the JMP "Barplot" display below. The default JMP Stemplot (**Stem and Leaf**) is also shown. Note that this differs from the one we created in that the Stem units are "split" and go from bottom to top. (So rotating has the x-axis reversed!)

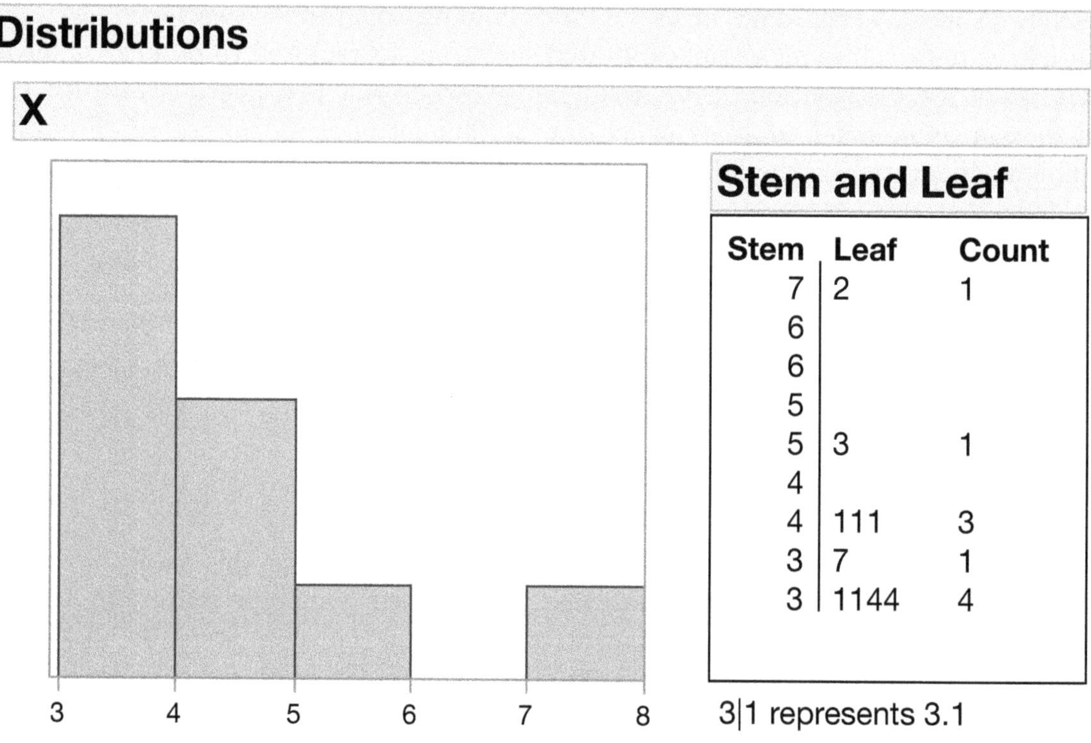

FIGURE 3.4. Barplot (1-line per Stem) and Stemplot (2-line per Stem)

Interpreting the Stemplot

Do NOT use a microscope — we are trying to assess the underlying distribution of the population based on a sample drawn from this population. That is, what is the general "pattern" of the Stemplot *profile,* and are there any obvious deviations from this profile. This will be discussed in more detail with respect to the Histogram, so for now, we just *let the display speak for itself.* What do we see in the "Barplot" and Stemplot displays?

Stemplot Profile

It is clear that the bar heights steadily decrease from 3 to 5 — there is gap at 6 — and the bar height at 7 is the same as at 5. Is the gap at 6 important? If it was, then it suggests we would *never* observe a data value at 6 in *any* new sample from the same population. Given that we have the data value 7.2 and the sample size $n = 10$ is small, this "constraint" does not make much sense. Note that if we were to draw a line connecting the center points at the top of each bar, the line would steadily decrease as we move to the **right**. This is indicative of what we call *right skew.*

Split Stemplot

Three variations of Stemplot are available for any given dataset. They are

```
1-line per Stem          2-line per Stem          5-line per Stem

Stem |  Leaf             Stem |  Leaf             Stem |  Leaf
-----|------------       -----|------------       -----|------------
   3 | [ 0 -- 9 ]           3 | [ 5 -- 9 ]           3 | [ 8  9 ]
                            3 | [ 0 -- 4 ]           3 | [ 6  7 ]
                                                     3 | [ 4  5 ]
                                                     3 | [ 2  3 ]
                                                     3 | [ 0  1 ]
```

That is, the Leaf digits [0 – 9] all go on 1 line, or are split over 2 or 5 lines per Stem unit. So which Stemplot display should we use? Equivalently, how did JMP decide on a 2-line per Stem format? This is quite a difficult problem since there is an inherent subjectiveness in the choice of display. Namely, our personal preference plays a part. However, JMP obviously uses some *rule* to decide, and the following guideline works well in general.

Stemplot Format Guide

Let $\{x_1, x_2, \ldots, x_n\}$ be the observed data values. Define:

$R = \max - \min$	The range of the data
$L = [10\log_{10}(n)]$	The maximum total number Stemplot lines[1]
$\Delta^* = R/L$	The initial "split" format

For the dataset in Figure 3.4 $R = 7.2 - 3.1 = 4.1$, $L = [10\log_{10}(10)] = 10$ so $\Delta^* = 0.41$

[1] $\log_{10}$ = log base 10 and $[y]$ = integer part of y e.g., $[2] = 2$, $[2.7] = 2$

Since the actual Stemplot format can only increase by 10 , 5, or 2 Leaf digits per line, we need to round up Δ^* to the nearest $10^r, 5 \times 10^r$, or 2×10^r for some integer r Hence 0.41 is closest to $0.5 = 5 \times 10^{-1}$ which suggests the 2-line display produced by JMP is preferred to our 1-line display.

Stemplot Notes

- Stemplots employ the actual data values in the display so they are really only useful for small to moderate datasets: $10 \leq n < 50$ (by hand) to 100 (using JMP).

- If two Stemplot formats are both reasonable choices, the display with more lines is usually recommended since this provides more resolution or detail about the Shape of the distribution.

  However, stretching out the display too much can introduce artifacts such as gaps and multiple lines having only 1 or 2 Leaf digits on them. These "artifacts" are not necessarily real features of the underlying distribution.

- The data values may first need to be rounded or truncated in order to create a suitable Stemplot display. For example, if $n = 11$ data values ranged from 10 to 417, then a 1-line per Stem display would require 41 lines!

  Truncation (417→41) has the slight advantage in that you "know" the dataset contains a data value 41x whereas rounding (417→42) could be 41x or 42x. In terms of the Stemplot display, it doesn't really matter which method is used.

- Since a Stemplot uses the actual data values multiplicities can be detected. For example, if you are given pulse rate data (heart beats per minute) and the Stemplot shows all the entries are multiples of 4, then it is clear the pulse rates of the patients were only measured for 15 seconds.

3.2.2 Histogram

We can use the same guidelines that were discussed for the Stemplot $\Delta^* = R/L$ save that we are *not* restricted to the three Stemplot formats with the Histogram display. In practice, we try and keep the "nice" formats (10, 5 or 2 $\times 10^r$) which are called "bin widths" for the Histogram display. We also have control over a *nice* starting value for the first bin.

We take the Frequency Table example that JMP used to create the 2-line per stem Stemplot in Figure 3.4 which corresponds to a Histogram with a bin width of $\Delta = 0.5$. What would be *nice* starting and ending "bin" values? Anything *below* $\min x_i$ will work so we start at $3 < \min\{x_i\} = 3.1$ and work up in bin width increments of $\Delta = 0.5$ until we exceed $\max\{x_i\} = 7.2$.

Class	Tally	Frequency	Proportion
3.0 to 3.5$^-$	////	4	0.4
3.5 to 4.0$^-$	/	1	0.1
4.0 to 4.5$^-$	///	3	0.3
4.5 to 5.0$^-$		0	0.0
5.0 to 5.5$^-$	/	1	0.1
5.5 to 6.0$^-$		0	0.0
6.0 to 6.5$^-$		0	0.0
6.5 to 7.0$^-$		0	0.0
7.0 to 7.5$^-$	/	1	0.1
Total		10	1.0

The notation 3.0 to 3.5$^-$ means that we will put any x_i satisfying $3.0 \leq x_i < 3.5$ in that "class" or bin. In this case there were 4 observations. Although none of the x_i in this dataset fell on a class boundary, if we obtained a new observation $x_{\text{new}} = 4.0$ this would go in the *third* class 4.0 to 4.5$^-$ and not the second class. In general, the choice of class boundaries: $3.0 \leq x_i < 3.5$ versus $3.0 < x_i \leq 3.5$ will not matter. It just needs to be consistent for each of the classes. JMP uses the first version for the histogram display. We added $x_{\text{new}} = 4.0$ which is highlighted in the third bar in Figure 3.5.

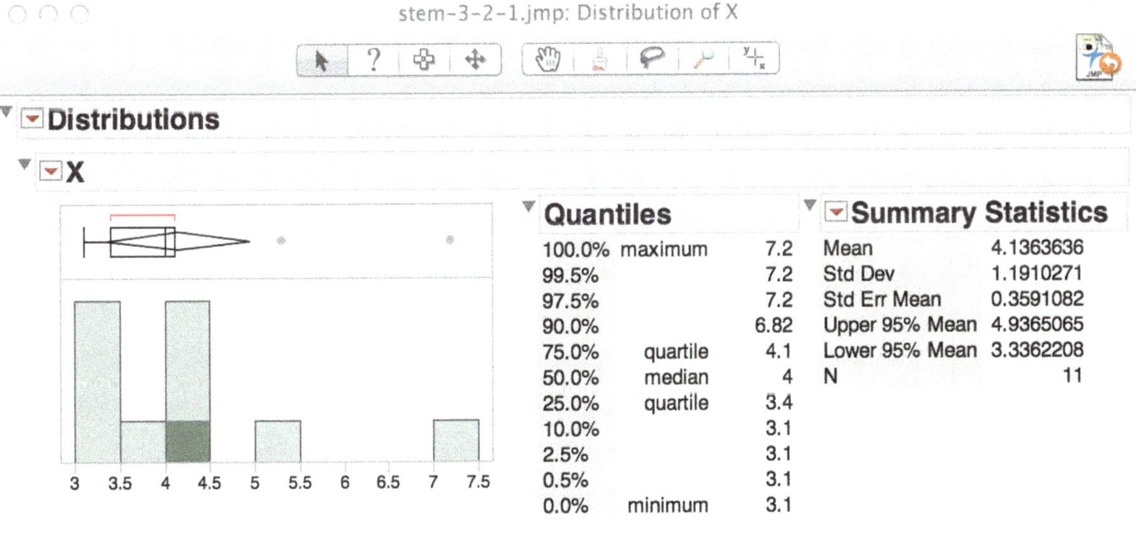

FIGURE 3.5. Histogram $\Delta = 0.5$

The default bin width used by JMP in Figure 3.5 can be modified by clicking on the "hand" icon and then on the histogram. Click-and-hold and moving up or down will make the bin width Δ smaller or larger respectively. Figure 3.6 (left) shows the result with $\Delta = 1.0$ and Figure 3.6 (right) with $\Delta = 2.0$ which clearly makes the resolution too coarse.

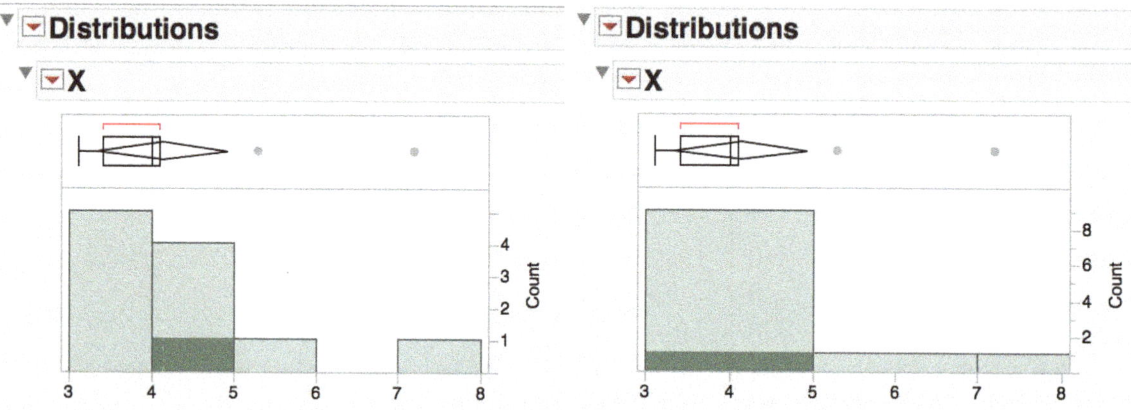

FIGURE 3.6. Histograms: $\Delta = 1.0$ and $\Delta = 2.0$

As we saw in the Stemplot display of Figure 3.4, there are gaps that are shown in the histogram display of Figure 3.5. The 2-line per stem and histogram in Figure 3.5 was the default choice made by JMP which agrees with the Stemplot Format Guide ($\Delta^* = 0.41 \nearrow 0.5$). However, this is only a guide and in this case, the 1-line per stem and histogram with $\Delta = 1.0$ provide a more realistic resolution of the **shape** profile where the bar heights decrease as we move to the *right.*

Key Point: Changing the *resolution* can enhance the interpretation of SHAPE

Interpretating the Histogram (and Stemplot)

- Look at the **overall pattern** of the shape profile — *what do we see?*
- Note any obvious **deviations** from the overall pattern.
- Describe the overall pattern in terms of **shape**, **location**, and **spread**.

Deviations:

Outliers individual values that lie far from where the majority of data values is located.

Clusters where several data values are grouped together at different locations.

Gaps places where there are no data values and so no bar appears (blank Stem unit).

Spikes bars that are unusually tall in comparison to neighboring bars.

Describing the Overall Pattern:

In describing the overall pattern, we try to assess the underlying distribution (**shape**) ignoring outliers or spikes. In addition, we would like to assign numeric measures for the **location** and **spread** exhibited by the data. For clusters, we describe each cluster separately. Outliers will obviously be separated by a large gap, whereas clusters may overlap. If there many small gaps, Δ may be too small (change the resolution), or there may be multiplicities in the data (e.g., pulse rates measured for 15 seconds then multiplied by four). We first introduce some terminology for describing the overall pattern.

Symmetric distribution where the profile of the bars on the left half of the histogram looks (approximately) like a mirror image of the profile on the right half.

Skewed distribution where one side of histogram "tails" off *consistently* over an extended range. Thus, we can have **right skew** or **left skew**

Location for symmetric distributions, this is the **center** of the horizontal scale (ignoring outliers). For other distributions, we can take it to be the location of the largest peak (also called the **mode**)

Spread is the range of the horizontal scale when no outliers or clusters are present. Ignoring outliers, this will be range of the distribution of the majority of data values. For clusters, it is the range of each cluster.

Examples of Deviations and Overall Patterns are shown in Figures 3.7, 3.8 and 3.9. To illustrate the "underlying" distribution, density (smooth) red curves have been overlaid on the histogram displays. For each plot, $n = 500$ simulated observations were used.

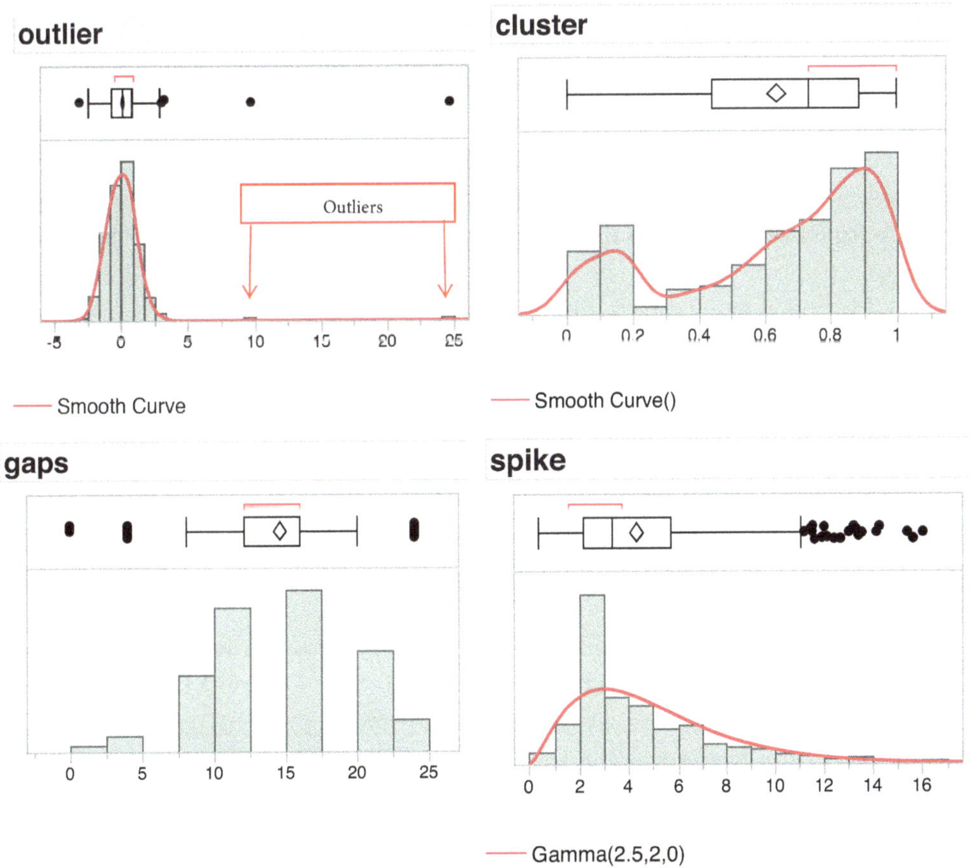

FIGURE 3.7. Deviations

Deviations

We refer to the four plots shown in Figure 3.7.

outlier plot shows two high values. Ignoring these values, the remaining data values appear to be symmetrically distributed over the range -3 to 3 and center close to 0. We use the term "outlier" to denote individual data values that seem completely "aberrant" to the distribution of the majority of data. Note that outliers are *not* necessarily "errors" and can be of significant importance. For example, the payout for someone who wins a lottery jackpot versus the payouts for all others who won something.

cluster plot shows the upper cluster is left-skewed over the range 0.2 to 1. The center as defined by the largest bar is close to 1. We can clearly see the lower cluster, but the resolution is too coarse to say anything useful about its shape. However, we can say the range seems to be confined to 0 to 0.2, and a rough estimate for center is around 0.1. The key issue with clusters is that the mechanism producing the data differs between the distinct cluster groups. For example, life expectancy for subjects with particular disease at birth may be significantly shorter than the natural life expectancy of subjects without that disease.

gaps plot illustrates a situation which would be considered very unusual. We have many (recall that $n = 500$) observations between 10 to 12.5 and 15 to 17.5, but apparently "zero" observations in between, and again no observations from 17.5 to 20. In addition, there is gap between 5 and 7.5. "Persistency" of missing observations (gaps) may be indicative of *multiplicity* in the data. With the data we can change the resolution to detect multiplicity, but without the data experience tells us that we need to investigate this further. That is, we go back to client and ask specifically how the data was collected (which we should have asked in the first place!)

spike plot shows the bar between 2 to 3 is substantially higher than the underlying right-skewed distribution would suggest for the remainder of the data values. Again this needs to be investigated by context, or with the client — "why is there an excessive number of data values between 2 and 3?"

Symmetric Distributions

Even with $n = 500$ simulated data values, the resulting histograms for the symmetric distributions in Figure 3.8 are hardly perfect. Hence, the caution **Do NOT use a microscope** advocated in the Stemplot interpretation. If the the left side of the histogram profile roughly matches the right side, then assuming the underlying (population) distribution is symmetric would be reasonable.

Note that we could also interpret the U-shaped distribution as two "clusters" so the context of the dataset is very important. For example, fuel efficiency in cars is optimal at highway cruising speeds. At both lower and higher speeds, fuel consumption increases. In this case, the U-shaped mechanism is consistent across all speeds and not due to cluster "groups."

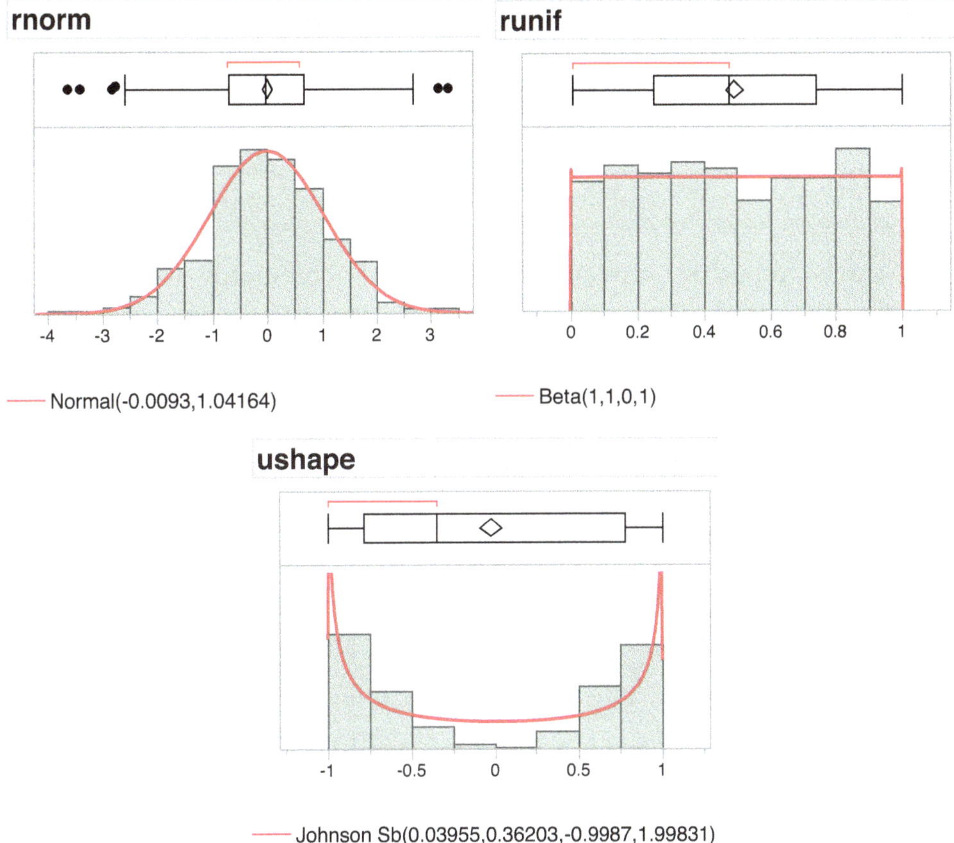

FIGURE 3.8. Symmetric Distributions

Skewed Distributions

Figure 3.9 illustrates the two forms of skewness. Thus, a skewed distribution is where the left or right side of the histogram "tails" off in a consistent manner over an extended range.

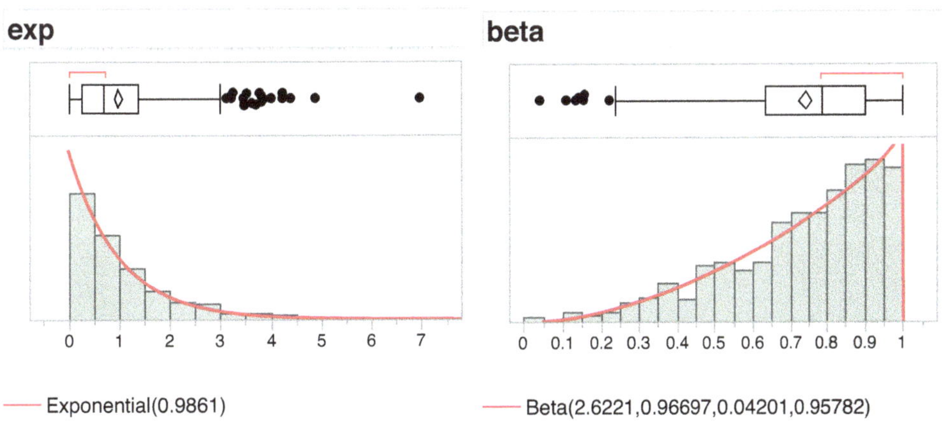

FIGURE 3.9. Skewed Distributions: **exp** = Right skew **beta** = Left skew

Note that in *strongly* skewed distributions there are often gaps that increase toward the end of the tail as the **exp** histogram. However, this is consistent with the underlying mechanism associated with skewness and these data values should not be regarded as "outliers" in the sense of being aberrant values as in the **outlier** histogram of Figure 3.7. The spike histogram (ignoring the spike) in Figure 3.7 is another example of a right-skewed distribution.

Summary

We now have a good repertoire of terminology that is commonly used to describe the **SHAPE** of the distribution of a dataset. However, our numerical "tools" for describing the **LOCATION** and **SPREAD** of a quantitative dataset is rather crude and pretty subjective when it comes to specifying the "center" of skewed distributions. We develop our numerical "toolkit" of summary statistics in the next section.

3.3 Location and Spread

We have already seen two numerical quantities that measure the LOCATION of a dataset, $\{x_1, x_2, \ldots, x_n\}$ say, in terms of the **extremes** (smallest and largest data values):

$$\textbf{min} = \textit{minimum} = \min\{x_i\} \quad \text{and} \quad \textbf{max} = \textit{maximum} = \max\{x_i\}$$

We also used these **summary statistics** to describe the SPREAD in terms of the "range" of the dataset as being from "min to max." Technically, this is **not** a "proper" measure of SPREAD since the **range** is a *single* number formally defined as:

range = **max** - **min** = a single numerical quantity

This might seem like semantics, but suppose we were **only** informed that *"There was a 10 point range on the midterm"* for scores on an exam, but we do not have the data. What were the **min** and **max** scores? We obviously have no way of knowing. Similarly, if we were **only** told the lowest score was 23 we cannot determine the range of scores. Of course, given both the min and max scores immediately provides more information such as the **range** and the **midrange** = $\frac{1}{2}(\min + \max)$ as a potential measure of the **center**.

The key point we are making is that in order to "analyze" a dataset using numerical summary statistics, the measures of LOCATION and SPREAD need to provide *independent* information about the dataset. [*Would you use your* `username` *as your* `password`*!?*]

We note that the min, max, and range do **not** provide any information about SHAPE. Hence SHAPE is another independent component of our EDA which we emphasized at the beginning of this chapter. Furthermore, these summary statistics can be quite misleading measures of location and spread as in the case of the **outlier** histogram in Figure 3.7. With the high outlier the range ≈ 28 is four time larger than the range ≈ 6 without both outliers. Expanding our toolkit of numerical summary statistics is needed.

3.3.1 Measures of Central Tendency

Mean

Whenever reference is made to the *average* value of a variable X such as SAT scores, height, etc., we are referring to the **mean** of X, or more precisely, the *sample mean* which we denote as $\overline{x}$ and pronounce as **"x-bar"** This is the most common measure used to describe where the **center** of a dataset is located. This is because we intuitively expect most of the data values will be around the average and fewer observations that have values which are substantially above or below average. The **mean** is defined as:

$$\overline{x} = \frac{\text{Sum up all the data values}}{\text{number of data values} = \text{sample size} = n} = \frac{x_1 + x_2 + \cdots + x_n}{n} \tag{3.1}$$

which can be more concisely expressed using the summation operator Σ as

$$\overline{x} = \frac{\sum_{i=1}^{n} x_i}{n} = \frac{1}{n}\sum_{i=1}^{n} x_i \quad \text{or just} \quad \frac{1}{n}\sum x_i \tag{3.2}$$

Thus $\sum x_i$ is understood to be the operation: *sum up all the data values in the dataset.*

Example 3.1 Let $X = \{1,\ 1,\ 4\}$ The mean of X is $\overline{x} = (1 + 1 + 4)/(n = 3) = 6/3 = 2$

Arithmetically, $\overline{x}$ is a special case of a summation wherein each value x_i is multiplied by the same *weight* $= 1/n$ Note that $\frac{1}{n}\sum x_i = \frac{1}{n}x_1 + \frac{1}{n}x_2 + \cdots + \frac{1}{n}x_n$ Statistically, we can interpret this in terms of how $\overline{x}$ *summarizes the information* in the dataset.

Key Point: $\overline{x}$ treats each data value x_i as *equally* important.

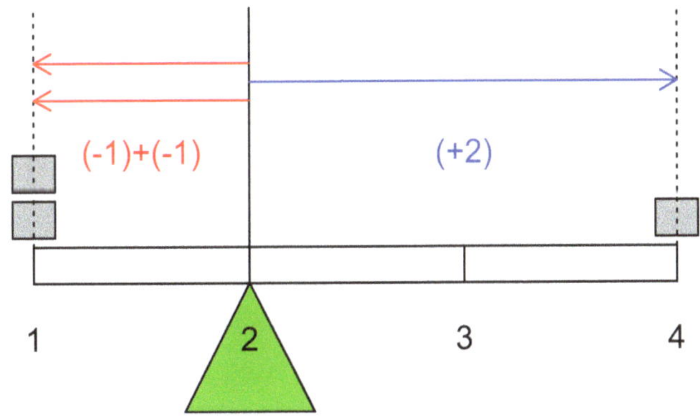

FIGURE 3.10. Balance point at $\overline{x} = 2$ (Green Pivot)

Another perspective is provided by computing the *deviations* $= (x_i - \overline{x})$ and summing up. For Figure 3.10 $\sum(x_i - \overline{x}) = (1-2) + (1-2) + (4-2) = (-1) + (-1) + (2) = 0$ Indeed, for any quantitative dataset it is *always* true that $\sum(x_i - \overline{x}) = 0$ and so $\overline{x}$ can be represented as the **balance point** which is illustrated by the physical scale in Figure 3.10.

Example 3.2 $X = \{67, 69, 71, 73, 220\}$ represent salaries (in \$1000s) at a small business.

Then $\overline{x} = 100$ so the "average" salary of workers at this company is \$100,000 (!) which would be quite an attractive marketing tool for hiring new employees. And this is **not** false advertising[2]. Of course, claiming the average salary is \$100K is very misleading since none of the current employees earn anything close to this value.

Key Point: We need more than just $\overline{x}$ in our numerical toolkit

Properties of the Mean

1. $\overline{x}$ is the **balance point** of a dataset. Synonymous terms used for physical objects (e.g., a vehicle) are *centroid* and *center-of-gravity.*

2. Outliers and skewed distributions will impact $\overline{x}$ since all the (sample) observations are given equal importance. Hence, $\overline{x}$ is said to be **nonresistant** to the influence of an individual data value.

Median

The **median**, which we denote as M, is the **middle** value of an *ordered* dataset. Thus, $M = 71$ for the salary example above. A problem arises when the sample size is *even* since technically the median can be *any* value that divides the ordered dataset into equal halves. For example, if a new hire accepted an offer of \$75K in the salary example above, then the ordered salaries are now $X = \{67, 69, 71, 73, 75, 220\}$ Thus, setting M to be **any** value between 71 and 73 (such as $M = 71.1$) would give the half datasets, $L = \{67, 69, 71\}$ and $U = \{73, 75, 220\}$, that have the same number of observations below and above "M" for any choice of M satisfying $71 < M < 73$. By *convention* we agree to define M as

$$\textbf{median} = M = \begin{cases} \text{middle value}, & \text{if } n \text{ odd} \\ \text{average of the two middle values}, & \text{if } n \text{ even} \end{cases}$$

Key Point: The *position* of M can be determined from $(n + 1)/2$

Example 3.3 $X = \{67, 69, 71, 73, 75, 220\}$ then $n = 6$ and $(n + 1)/2 = 3.5$ This tells us that we need to average the 3rd and 4th data values in the *ordered* dataset. Hence $M = (71 + 73)/2 = 72$. Without the new hire (75K), $n = 5$ and $(n + 1)/2 = 3 \Rightarrow M = 71$

With or without the new employee, we can see that M provides a measure of **center** that better reflects the *majority* of salaries at this company. Furthermore, M will remain the same no matter how much the CEO salary increases (assuming it is currently \$220K).

Mean versus Median

1. We refer to M as the **halfway point** of an ordered dataset since the number of data values that lie below M is the same as the number of data values that lie above M. It is also called the *50th percentile.* $\overline{x}$ is the average or **balance point** of a dataset.

[2] While it is true that $\overline{x} = 100$, the Legal context of "false marketing" this claim is a different matter.

2. M is more **resistant** to the influence of outliers and skewed distributions, whereas both these situations will have a direct impact on $\overline{x}$.

3. If the distribution is **symmetric** then $M \approx \overline{x}$ In skewed distributions $M < \overline{x}$ (right skew) and $\overline{x} < M$ (left skew). The converse does **not** necessarily hold. That is, $M \approx \overline{x}$ does not necessarily imply the distribution is symmetric.

4. M is impacted by **rounding**. For example, if salary data is rounded to the nearest \$1000, M will always be an actual salary or the average of two salaries, no matter how many employees a company has (i.e., $M = 70.2$ is not possible). The impact of rounding on $\overline{x}$ diminishes rapidly as n increases.

Question: So which measure of **center** should we use?

Key Point: Always, always, always plot the data FIRST

If the distribution is symmetric, then either measure, $\overline{x}$ or M, could be used, but it is standard practice to compute both for EDA. Outliers will impact the mean so they should be excluded before computing $\overline{x}$. For skewed distributions, the median is often the preferred measure. For example, *median income* and *median house prices* are reported by government statistics and used for comparison purposes.

The main issue with skewed data is that there is no *natural* center unlike symmetric distributions. That is, while the median may be the preferred measure to use for skewed data, it can still be quite misleading. A simple example is $X = \{1, 10, 100, 1000, 10000\}$ where neither $M = 100$ and $\overline{x} = 2222.2$ provide a realistic measure of a "typical" data value. Of course, if we *transform* the data using $Y = \log_{10} X = \{0, 1, 2, 3, 4\}$ we now have a symmetric distribution with a natural center $\overline{y} = M_y = 2$.

3.3.2 The 5 Number Summary

Let L = Lower half of the dataset (all the data values below M) and U = Upper half of the ordered dataset. Then we define the **quartiles** as:

$$\text{Lower quartile} = Q_1 = \text{med}\{L\} \quad \text{and} \quad \text{Upper quartile} = Q_3 = \text{med}\{U\}$$

Like the median, we can also refer to Q_1, Q_3 as the 25th and 75th percentiles, respectively.

Example 3.4 Let $X = \{$ $\boxed{\boxed{10,\ 15,\ 20,\ 28,\ 34},\ 44}$, 50, 80, 100, 295, 415$\}$

Here, $n = 11$ so M is the 6th data value = 44. The lower boxes show that we have *two* choices for L when n is **odd**: all the data values *strictly* below M (inside box) or include M as well (outside box). Thus, $Q_1 = 20$ without M, or $Q_1^* = (20+28)/2 = 24$ with M included. Unlike the rule for computing M from a sample dataset, there is no agreed upon *convention* for computing the quartiles. Hence, either choice is valid provided the same procedure is used to compute the upper quartile from U: $Q_3 = 100$ without M, or $Q_3^* = (80+100)/2 = 90$ with

M included. In practical terms, it makes little difference and is more a matter of convenience if we are computing the quartiles manually. Note that this only applies when n is odd since the L and U half-datasets are uniquely defined when n is even.

Key Point: The quartiles are just lower and upper *medians*. No additional procedure is required[3].

5 Number Summary (5NS) For a quantitative dataset: $5NS = \{\min, Q_1, M, Q_3, \max\}$

For the example above, $5NS = \{10, 20, 44, 100, 415\}$ [or $\{10, 24, 44, 90, 415\}$ if the median $M = 44$ is included in L and U] and can be graphically displayed as a **Boxplot** To create an effective Boxplot display, we need the following quantities:

Inter-Quartile Range $IQR = Q_3 - Q_1$

This is a measure of SPREAD (like the **range**), but for the *middle* 50% of the dataset. As such, it is resistant to outliers, and therefore can be used to define a criterion to *detect* potential outliers. The "cutoffs" suggested are:

Outlier Criteria: $C_L = Q_1 - 1.5 * IQR$ and $C_U = Q_3 + 1.5 * IQR$

Thus, $x_i < C_L$ or $x_i > C_U$ would suggest x_i is a potential outlier. The value 1.5 is based on trial-and-error simulations (and theory) and in general, provides a good compromise between detecting outliers versus being oversensitive. This criterion does **not** do well if the underlying distribution is strongly skewed.

3.3.3 Boxplot

A Boxplot is a graphical representation of the 5NS and provides a rather concise, but effective visualization of the Shape (distribution) of a dataset. Parallel Boxplots enable multiple groups to be compared in one display. The following steps are used to construct a Boxplot for $X = \{x_1, x_2, \ldots, x_n\}$

1. Compute the 5NS
2. Compute the outlier cutoffs C_L and C_U Note any data values that are potential outliers
3. Draw a true (calibrated) scale that spans the min and max of X
4. Use a nice starting point and increments (see Histogram discussion)
5. Draw the box left side = Q_1 , right side = Q_3 , M = Median (usually) lies within the box
6. Add the outliers (if any) as isolated points
7. Draw the "whiskers" to the data value *closest* to C_L and C_U that is not an outlier

   If there are no outliers, the whiskers go to the min and max (bottom left boxplot)

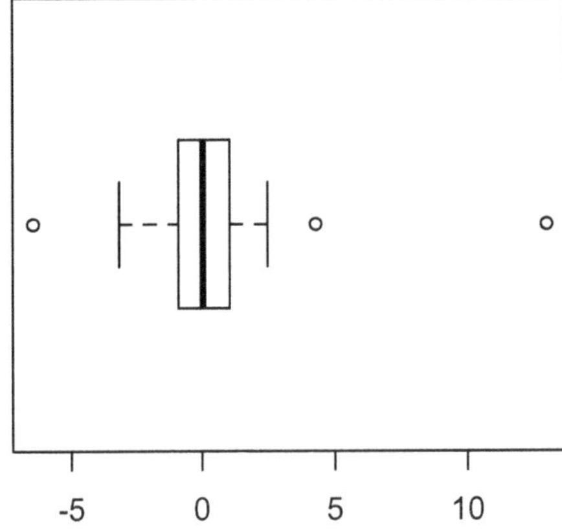

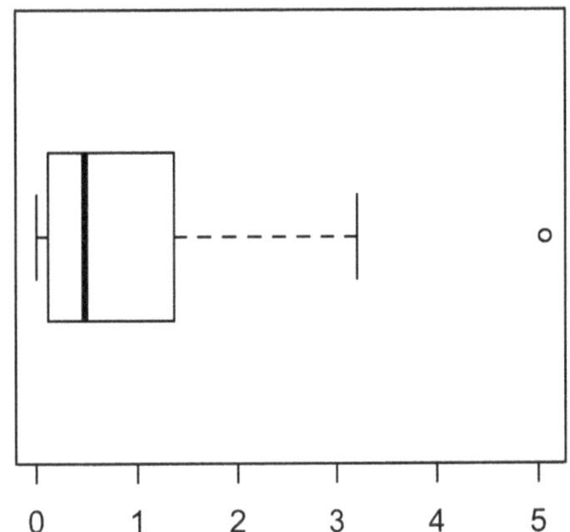

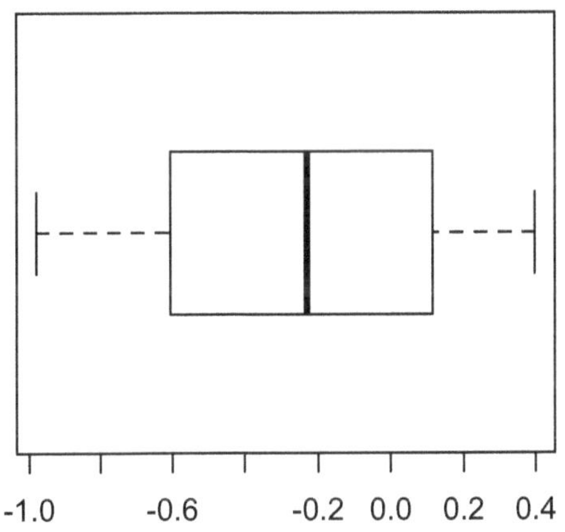

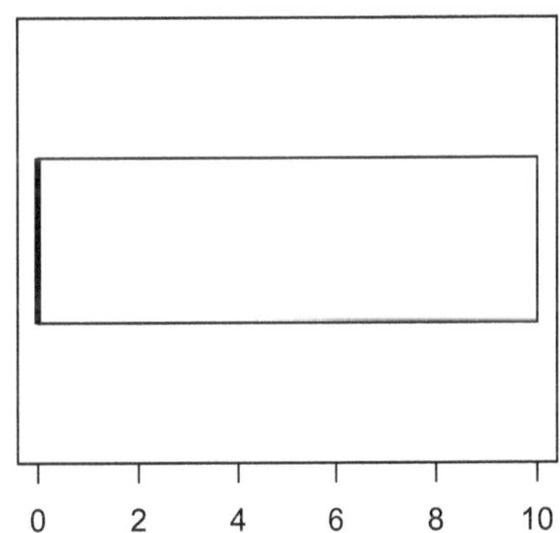

FIGURE 3.11. Boxplot Examples

In small datasets, the Boxplot display may look rather strange as in the bottom right boxplot. Here, we recorded precipitation for 10 days. Six days were sunshine (no rain), interspersed with four days of showers that each produced 0.1 inches of rain (recorded as "10") Hence, the (ordered) dataset is: $\{0, 0, 0, 0, 0, 0, 10, 10, 10, 10\}$ which gives the 5NS $= \{\min = Q_1 = M = 0,\ Q_3 = \max = 10\}$ Thus, the bottom right Boxplot display!

[3] JMP uses a weighted average for the quartiles, so the JMP quartiles may differ slightly

Interpretation

The Boxplot is a graphical display of a dataset in *25 percent increments.* Thus, we interpret the SHAPE of the underlying distribution by comparing the ranges of each increment. Since it based the 5NS, which a very reduced summary of a dataset, the Boxplot does not have the same resolution as a Histogram. However, it can be quite effective, particularly in comparing groups. For a single dataset the features of interest are:

- **Deviations** Outliers are depicted as isolated points which will will appear above and/or below the whiskers that extend from the box. A Boxplot cannot tell us if Clusters are present, so if we know the data contains "groups" we should use separate or parallel Boxplot displays.
- **Symmetry** Ignoring outliers, a symmetric distribution is suggested when **both** the following hold.
  1. The distance from M to the ends of the whiskers are approximately the same
  2. M divides the box into equal halves: $M - Q_1 \approx Q_3 - M$

  Note that if an outlier is close to the end of a whisker, we may be able to use it to assess symmetry. For example, if the right whisker was extended to the outlier in the top left Boxplot, it would be roughly the same length as the lower whisker. Hence, symmetry is suggested ignoring the other two outliers.
- **Skewness** The range of each successive increment should steadily increase in a right skewed distribution as in the top right Boxplot. Conversely, the ranges should steadily decrease in a left skewed display. Mild skewness would be suggested by a slightly longer whisker on one side as in the bottom left Boxplot.

Parallel Boxplots

Figure 3.12 shows parallel Boxplots for two types of NO emission detection devices. It can be seen that the `Type I` device detected much more NO than other device. We also see that the distribution of the `Type I` device appears slightly right skew (top half is more stretched out), whereas the `Type II` device appears to have symmetric distribution.

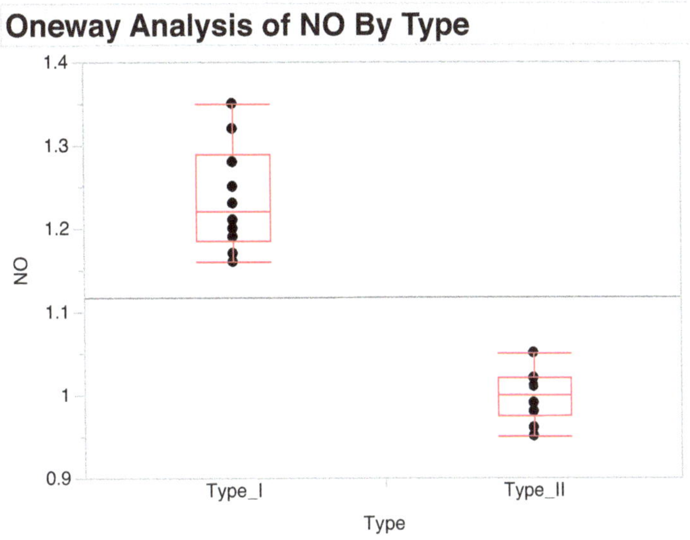

FIGURE 3.12. Parallel Boxplots

3.3.4 Standard Deviation

The standard deviation, denoted as s throughout this text, is the SPREAD *companion* to the LOCATION measure $\overline{x}$. That is, $\overline{x}$ and s provide independent information about a dataset with respect to its center and variability. It is defined by

Standard Deviation: $\texttt{stdev} = s = \sqrt{\frac{1}{n-1}\sum_{i=1}^{n}(x_i - \overline{x})^2}$

which seems rather complicated and certainly something we would *never* want to compute manually. We will leave the computation of numerical summary statistics to JMP.

Properties of Standard Deviation:

1. `stdev` ≥ 0 for any dataset.
2. `stdev` $= 0$ **only** if there is **no** variability $\Rightarrow x_i = c$ (constant) for all data values.
3. `stdev` has the same **units** as the original data so results can written as: $\overline{x} \pm s$
4. Since `stdev` uses $\overline{x}$, it is **not** resistant to outliers.

Remark: Even for small to moderate datasets ($n < 50$) computing $\overline{x}$ with a calculator is "barely" doable, but computing `stdev` would be an act of madness! Fortunately, we can obtain an **approximate** value of `stdev` using

$$\texttt{stdev} \approx \frac{\textbf{range}}{4}$$

which can be manually obtained from a stemplot (excluding outliers). Furthermore, if the stemplot is roughly symmetric, then we can take $\overline{x} \approx M$

Derivation of `stdev`

We would likely agree that the formula for `stdev` looks rather unintuitive, but there is a logical derivation which we present here.

The idea is quite simple: *using* $\overline{x}$ *as the* **center** *we want to derive a measure of variability (SPREAD) about the* **mean** $\Rightarrow$ we need to consider the **deviations** $(x_i - \overline{x})$ Hence, it might seem reasonable to use the "average deviation" as our (single) measure of variability.

Problem: Average deviation $= \dfrac{\text{Sum of deviations}}{\text{sample size} = n} = \dfrac{\sum(x_i - \overline{x})}{n} \equiv 0$ for ALL datasets

This is because $\overline{x}$ is the **balance point** so $\sum(x_i - \overline{x})$ always equals zero. Not useful at all ! Since $X = \{1,\ 1,\ 4\}$ clearly has smaller spread than $Y = \{10,\ 10,\ 40\}$ we expect our measure of variability to show this numerically i.e., `stdev`$(X) <$ `stdev`(Y)

The deviations for X are $\{-1,\ -1,\ 2\}$ and for Y $\{-10,\ -10,\ 20\}$ which are 10 times larger. Hence, strip the "sign" and use the *average deviation* **magnitude** $= \frac{1}{n}\sum|x_i - \overline{x}\,|$ which is called the **Absolute Deviation**. This is a completely valid measure of variability, but has the disadvantage that it is not easy to develop *inference* procedures (that we need in later chapters) based on the Absolute Deviation summary statistic for SPREAD.

The technical issue above can be avoided by using "squared" *deviations.* This just makes a large deviation, bigger, and also ensures that it will be positive. Hence, the *average* **squared** *deviation* $= \frac{1}{n}\sum(x_i - \overline{x}\,)^2$ is simply a numerical value that we can interpret in terms of variability: large or small value $\Rightarrow$ large or small SPREAD, respectively. Apart from the $(n-1)$ instead of n in the formula for `stdev` this is effectively s^2 So why $(n-1)$?

Consider $X = \{1,\ 1,\ ?\}$ where "?" is unknown, but we are told $\overline{x} = 4$ Then we know $(1{+}1{+}?\,)/(n=3) = \overline{x} = 4 \Rightarrow\ ? = 10$ That is, knowing $\overline{x}$ is "equivalent" to knowing a single data value (if we know the other $(n-1)$ data values). Another way to think of this is that in order to compute `stdev` we first need to compute $\overline{x}$ and so we have "used up" one (1) piece of information from the dataset. That leaves an *effective* sample size of $(n-1)$ which is also referred to as the **degrees-of-freedom** parameter. Hence, the reason for $(n-1)$ in the formula for `stdev`.

Sample Variance The quantity $s^2 = \frac{1}{n-1}\sum_{i=1}^{n}(x_i - \overline{x}\,)^2$ is called the sample variance.

Finally, let X be our daily travel distance in miles which we record for a particular month. Then $\overline{x}$ is the *average* daily distance (in miles) we travelled, and $(x_i - \overline{x})$ is the daily deviation (also in miles) from our average. But, s^2 would be an overall measure of our "average" variability — in **miles**^{2} — and we would be saying *"Our average daily travel was 25.5 miles* $\pm$ *225* **square miles** ! Taking the square root gives `stdev` $= 15$ miles which now has the same *units* as the original data values.

3.3.5 Which Measures ?

Our toolkit now includes several graphical and numerical summary statistics. Since JMP will give us all of these EDA diagnostic measures, the question is more about which set of measures are better suited for describing the distribution of a particular dataset.

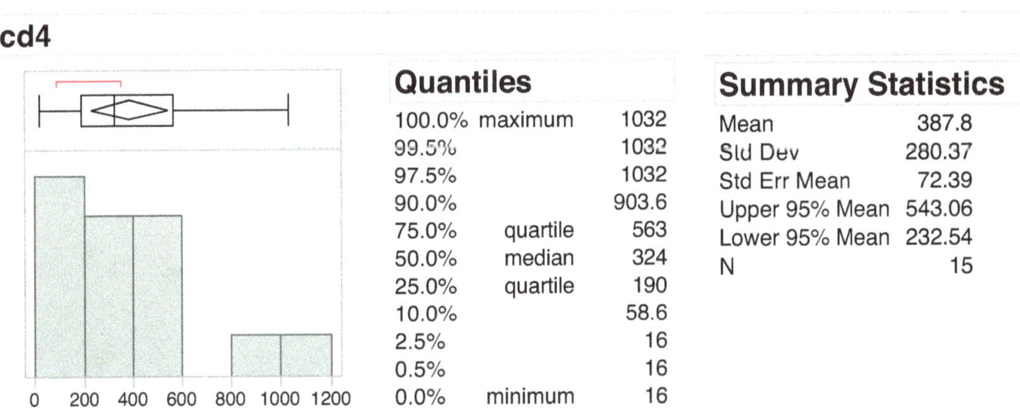

FIGURE 3.13. CD4 Dataset: JMP Output

The standard trio is: **Histogram, Mean, Standard Deviation** This is the preferred set of summary statistics when the underlying distribution is close to symmetric. When the distribution is clearly skewed, the **5NS** provides a more representative description of the dataset and the Boxplot provides a visually less "cluttered" overview of the skewness. Figure 3.13 shows the default output from JMP for a CD4 dataset of HIV infected patients. [CD4 are white blood cells that fight infection (also called T-cells). A normal count range is between 500 to 1500; below 500 means your immune system is compromised.]

3.3.6 Timeplots

A **time series** is a special case of a dataset where the data values are collected *sequentially* in time. This means the **order** of the the recorded data values is important. For example, if y_t is today's high temperature, then y_{t-1} was yesterday's high and y_{t+1} will be tomorrow's high (which we won't know precisely until tomorrow). Hence the 7-day weather forecast, tracking daily positive COVID-19 cases, daily Dow Jones Index stock market close, monthly job creation, electrocardiograms (EKG), major earthquake events, are applications involving time series data which can be graphically displayed as a **timeplot**.

Figure 3.14 shows monthly US housing starts (in 1000s). To enhance the visualization of the timeplot to help detect **trends** and **cyclic** behavior, we connect the monthly data values with the red line. As expected, fewer housing construction starts will occur during the winter months versus the summer months. (This is not shown by the `Row` variable, but we can deduce this from the *context* of the data.) So we see a strong yearly cyclic pattern that fluctuates; sometimes decreasing quickly from the peak for some years, with other years staying close to the peak for several months. We can also see there was "level" shift upward

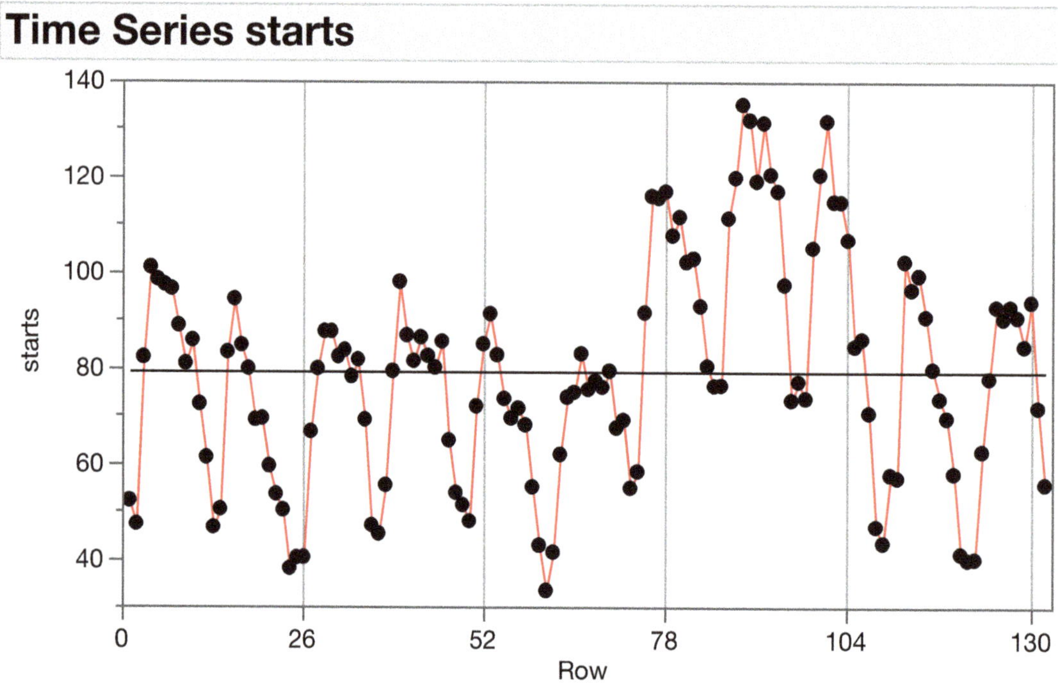

FIGURE 3.14. Monthly Housing Starts: JMP Output

in years 7, 8, 9 whereas the other years were more-or-less flat (in terms of average yearly starts). This is the trend component of the time series.

Key Point: We interpret a timeplot by describing **obvious** trends and cyclic patterns.

Remark: Figure 3.14 illustrates the situation where the time series is observed on a *regular* monthly basis. Similarly for the daily examples mentioned above. However, a time series can be continuous as in the EKG example (your heart does not stop working between beats!), or irregularly spaced such as in major earthquakes events. Lastly, the time *index* of a time series also applies to **spatial** data. For example, the number of broken-down cars on a highway can be indexed by mile markers.

3.4 JMP Instructions

This chapter contains many examples of graphical and numerical output from JMP. Since JMP is menu-driven, creating the output requires several steps as shown in Figure 3.15 Appendix A provides a more detailed overview of JMP which you may find useful to refer to as we describe the JMP output process for producing the `gender` barplot in Figure 3.1

Barplot: The JMP dataset "could" simply have a `gender` column with 20 rows indicating the gender, `Male` or `Female`, of each person. For categorical variables, we also have the option of providing JMP with the *total* frequency of each group which is labelled as **Kount** that we use to illustrate the functionality of the **Dialog** window. Here is `mydata.jmp`

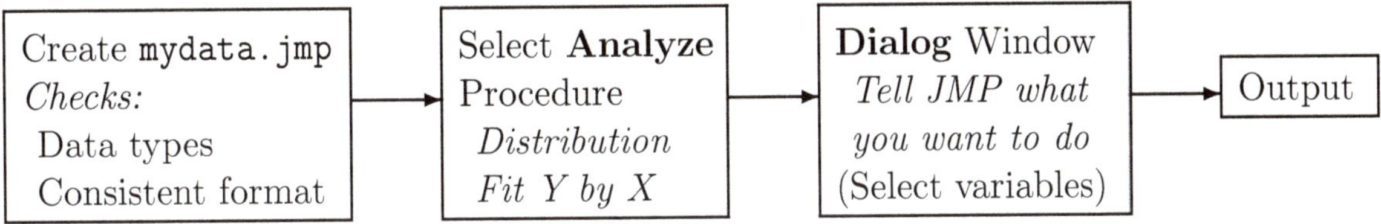

FIGURE 3.15. Basic JMP Output Steps

	Gender	Kount
1	Female	8
2	Male	12

Now select **Analyze** → **Distribution** and complete the **Dialog** window. Note that we need to tell JMP to use `Kount` as the Freq count for `Gender`. Then click OK

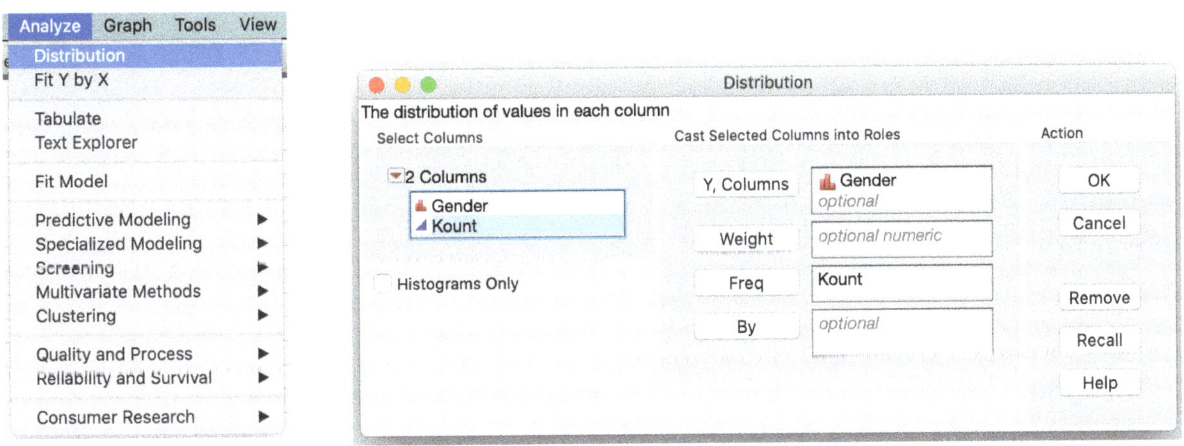

JMP produces a *vertical* display (left) by default which we can change to a *horizontal* layout (right) by selecting **Distributions** → **Stack**. Further adjustments can be made to the barplot indicated by the blue (right side) and red (lower right corner) circles. The mouse pointer needs to be *delicately* placed on these positions. Then click-and-hold.

Blue circle: (cursor changes to ⇹) Move left or right to shrink or expand the *width* of the barplot. Red circle: (cursor changes to ⤡) Resize barplot height and width as desired.

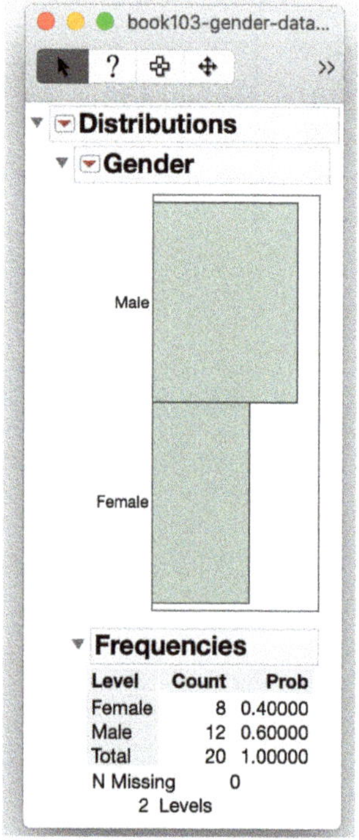

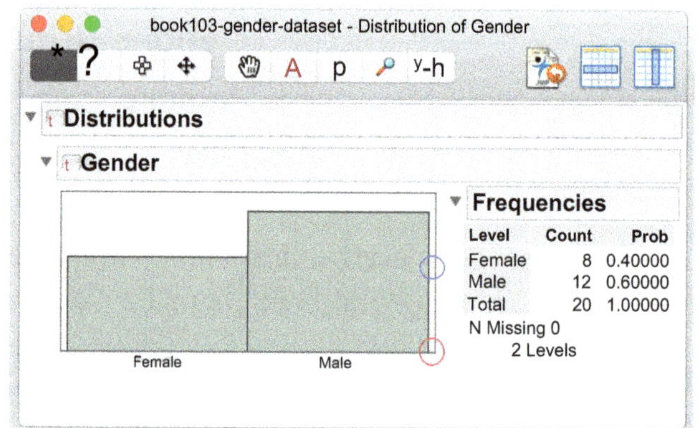

Histogram: For a quantitative variable X all the data values are entered in the `X` column (no **Kount** column). Then do **Analyze** → **Distribution** and add `X` to the Y, Columns box in the **Dialog** window. The default Output includes the Histogram, Quantiles and Summary Statistics reports as shown in Figure 3.13

Timeplot: Let Y be a time series. Creating a timeplot such as Figure 3.14 can be produced in several ways. We describe three methods for *regular* Y (equally spaced time intervals).

1. **Analyze** → **Specialized Modeling** → **Time Series** (submenu)

   In the **Dialog** window, enter `Y` in the Y, Time Series box. Then click OK

   This uses the **Row** number as the default *Time* index (see Figure 3.14)

2. **Graph** → **Legacy** → **Overlay Plot**

   In the **Dialog** window, enter `Y` in the Y box. Then click OK

   In the Output window (graph) select **Show Control Panel**

   Then click on the 6th icon to connect the points.

3. **Analyze** → **Fit Y by X**

   We first need to create a new **Time** column (this is `X`) in the time series JMP dataset.

   (a) Select **Cols** → **New Columns...** Rename as `Time`

   (b) Click on the `Time` header (highlights the column). Select **Cols** → **Formula...**

(c) In the new popup window, select **Row** (left panel) → **Row** Then click OK
[This puts the **Row** number as the `Time` entries]

Now do **Analyze** → **Fit Y by X** and put `Y` in the Y, Response box and `Time` in the X, Factor box. Click OK

Finally, in the Output window (graph) select the option **Flexible** → **Fit Each Value** to connect the points

Parallel Boxplots: When X is a categorical variable and Y is a quantitative variable, then **Analyze** → **Fit Y by X** will produce a parallel Boxplot display such as Figure 3.12

To get the actual Boxplot(s), select **Display Options** → **Boxplots**

Remark: Our primary statistical analysis platforms will be **Analyze** → **Distribution** (univariate X) and **Analyze** → **Fit Y by X** (bivariate X, Y)
Instructions for creating JMP output will be included here and in subsequent chapters.

Chapter 3 Exercises

NOTE: We assume JMP will be used to create output needed for the Exercises.

3.1 Suppose we estimate missing values of a quantitative variable X by using the *average* of the non-missing responses. That is, set all $x_{\text{miss}} = \overline{x}_{\text{nomiss}}$
Does this seem to be a reasonable procedure (or not)? Explain. [HINT: `stdev`]

3.2 PULSE The heartbeats per minute (pulse rate) for 39 individuals is shown below.

```
52 56 60 60 60 60 60 60 64 64 64 64 64
64 64 68 68 68 68 72 72 72 72 72 72 72
72 74 76 76 76 76 80 80 84 88 88 88 92
```

a) Create a histogram and stemplot. Describe the distribution.

b) Use the "hand" icon to reduce the bin-width of the histogram display to the frequency of the individual values. What does this show? Be specific.

3.3 OLDFAITH This dataset contains several variables associated with the geyser "Old Faithful" located in Yellowstone National Park. In particular, Duration = Length of the eruption in minutes, and Height = Estimated maximum height of the eruption in feet.

a) Explain **why** Duration shows two distinct "groups" in the histogram display
To answer the "**why**" we first need to research *how does a geyser erupt?*

b) Does Duration have an impact on Height? Explain.
Plot the histogram of Height. In the Duration histogram, drag (click-and-hold) the cursor over the long Duration bars. This will highlight the corresponding Height values. Do the same for the short Duration bars. Is there any obvious difference?

3.4 A researcher claimed they had a dataset consisting of four observations where two were greater than two, but the sample mean was zero and the variance was four. (`stdev` = 2). Show that if $\overline{x} = 0$ then `stdev` cannot be 2 (indeed $s > 2$)

3.5 To illustrate the Boxplot display, a statistics professor employed the data below

`10 15 20 28 34 44 61 78 110 289 417`

a) i) Compute the five number summary for these data.
ii) Calculate the interquartile range.

b) Hence sketch an *appropriate* Boxplot display.

c) What problem arises if we wanted to create a Stemplot display? Be specific.

3.6 Sketch a barplot or histogram that would represent the following:

1. Left-handed and right-handed students in a large introductory college course
2. The gender of the students in the course
3. The height in inches of the students in the course
4. Time spent on their phones daily (in minutes on average)

3.7 SSHA The Survey of Study Habits and Attitudes (SSHA) is a psychological test that measures students attitudes toward school and study habits. Scores range from 0 to 200 (the highest positive attitude score). The dataset are scores for 38 first-year college students.

a) Create separate stemplots for the Female and Male scores.

b) Compute the five-number summary for the 19 Females.

c) Is the high SSHA Female score a potential outlier? Explain.

d) For both genders, where will the median lie in relation to the mean?
[You do **not** need to compute M or $\overline{x}$ to answer this.]

3.8 THERAPY Two therapies for patients with severe heart disease. Dataset gives the survival times (months) under the two therapies.

a) Construct separate histograms for the two therapies

b) Discuss and compare the survival times of the two therapies.

3.9 SAT Math and Verbal SAT Scores by Gender (1967–1996) are given in the dataset

a) Plot the four time series on the same graph. Do the following:
Fit Y By X Enter X =Year, Y = Score
Select **Group By ...** enter GT Then **Flexible** → **Fit Each Value**

b) Discuss the trends for Males versus Females and Math versus Verbal SAT scores

4
Probability Distributions

The EDA methods discussed in chapter 3 apply to the observed sample. However, the smooth curves shown in Figures 3.7, 3.8 and 3.9 suggest the **notion** of an underlying distribution associated with the **population** from which the sample was selected. It follows that "if" we could measure **every** item or subject in the population, we could calculate the **true** mean.

Since $\overline{x}$ was computed from a sample which is only part of the population, we cannot expect $\overline{x}$ to be equal to the true mean. Thus, we make a distinction between quantities associated with a sample versus quantities associated with the population from which the sample was selected.

Statistic Any quantity that can be computed from a sample.

Parameter A quantity associated with the population

Greek letters are used to distinguish population parameters, such as μ, σ for the population mean and population standard deviation. In practice, a population parameter will be **unknown** since it is too expensive, impractical, or even impossible to measure every item in the population (section 2.1). So can we say anything useful about the *population* based on a sample?

Question *Given $\overline{x}$, what can we say about μ?*

This is called **inferential statistics** and is an important component of Statistical Analysis wherein we make **conclusions** about a population parameter of interest based on sample statistics. We will address this in more detail in the following chapters.

Question *Why do we want to make conclusions about an unknown population parameter?*

Since the parameter is unknown, surely any conclusion would be meaningless as we will never know if the conclusion is true ... That is absolutely correct ! We can never be "certain" our conclusion is true. This would therefore seem to be a rather pointless exercise.

A counterexample is clinical trials. A new drug is tested (such as the COVID-19 vaccines) and based on the sample evidence, it is concluded the **entire** target population will benefit. The COVID-19 vaccines were not perfect — some patients had adverse reactions or did not respond to the treatment. But over 90% of the study patients benefitted with reduced symptoms or not having to be hospitalized. Hence, it was concluded that getting the general population vaccinated would substantially help mitigate the COVID-19 pandemic.

4.1 The Model Paradigm

$$\text{DATA} = \text{MODEL} + \text{ERROR} \tag{4.1}$$

Throughout the remainder of this book, we assume that the purpose of the statistical analysis is to **summarize** the DATA in terms of a MODEL which captures the **structural** features of interest, leaving the ERROR component which defines the inherent **randomness** associated with the process. For example, the *population* mean model can be defined as

$$x_i = \mu + \boxed{\epsilon_i} \tag{4.2}$$

where x_i is the observed value, μ is the (unknown) true population mean, and ϵ_i represents the (unknown) **random** deviation of x_i from μ. From our EDA knowledge, we could *estimate* the population mean model in Equation 4.2 as

$$x_i = \bar{x} + e_i \tag{4.3}$$

where we could use the **residuals** e_i from Equation 4.3 to provide an estimate (histogram) of the underlying probability distribution of the inherent randomness associated with the process. But this will just be a sample estimate of the probability distribution of the ERROR term in Equation 4.1

It follows that for us to transition to the population model represented in Equation 4.2 we will need to make assumptions about the probability distribution of $\boxed{\epsilon_i}$ Hence we need to understand **probability**.

4.2 Probability

Probability is the study of random phenomena. In the simplest case probability attempts to address the following question in mathematical terms

> *"What is the chance of Event A occurring?"*

Two approaches are:

Frequency (Empirical): Given A is the event of interest, the above question can be answered by performing an experiment N times and assigning the measure $P_N[A] = N_A/N$, where N_A is the number of times the event A occurred.

Axiomatic (Theoretical): The measure $P[A]$ can also be assigned under the intuitive notion that there is some inherent chance associated with the event A. In this abstract approach, a set of probability rules called *axioms* are imposed which we discuss later.

Example 4.1 Coin Toss The presumption that one has a 50-50 chance of getting Heads prior to any toss corresponds to (*axiomatically*) assigning $P[H] = 0.5$. This probability measure seems reasonable (or obvious) given that the only possible outcomes are H or T.

- If the coin is fair, we would expect $P_N(H) = N_H/N$ to **converge** (get closer) to the value 0.5 as N gets larger.
- However, if the coin is biased toward H or T, then $P_N(H)$ would converge to a different value than was "assumed" under the axiomatic approach.

Key Point: Empirical evidence is needed when the Axiomatic approach is unknown.

For example, medical risk factors are based on empirical evidence. What is the recommended age that women should start to get mammograms, or at what age should men get prostrate exams. These guidelines are **not** determined on an "assumed" *axiomatic* basis.

4.3 Set Theory and Probability Rules

Probability consists of two components:

- the **event** A
- the probability measure $P[A]$ assigned to the event A

The difficulty comes when we have **composite** events, A and B (and C or more). We start with a simple example

Example 4.2 Fair Die Toss Let the events of interest be $A = \{2, 4, 6\}$ and $B = \{1, 2, 3\}$ Then $P[A] = P[B] = \frac{3}{6} = 0.5$ is certainly reasonable. But, what about the probability of the *composite* event $A``+"B$?

We would agree that $P[A``+"B] \neq P[A] + P[B] = 1$

since this implies the event $A``+"B$ is *certain* to occur which cannot be true as the outcome $\{5\}$ is missing and $P[\{5\}] = \frac{1}{6}$

The issue here is that the operation $A``+"B$ between events is **not** the same as adding two real numbers $P[A] + P[B]$ Furthermore, $A``+"B$ has two distinct interpretations:

All the outcomes in A **or** B (or both) = {1, 2, 3, 4, 6} ; Only the outcomes that are common to A **and** B = {2}

To resolve this situation, we employ set operators.

4.3.1 Set Theory

To develop the Axiomatic approach, events need to be described in precise terms. This can done using set theory.

Special sets:

$\mathcal{S}$ = Sample space. Contains all possible outcomes.
$\emptyset$ = Empty set. The null or impossible event.

Set Operators:

We can use the three operators below to describe any composite event of interest.

$\cup$ = union. All the outcomes from combining two events.
$A \cup B$ is the event where A **or** B (or both) occur.

$\cap$ = intersection. The outcomes are common to two events.
$A \cap B$ is the event where both A **and** B occur.

$\overline{E}$ = complement (of event E). The event of **not** getting E.

Disjoint Events:

Events A and B are disjoint if $A \cap B = \emptyset$ (no outcomes common to A and B)
(Also called *mutually exclusive*)

4.3.2 Probability Rules

We now define the three axioms of probability

Probability Axioms:

1. For any event A, $0 \leq P[A] \leq 1$
2. The sample space is certain to occur. $P[\mathcal{S}] = 1$
3. If the events A_1, A_2, $A_3 \ldots$ are disjoint, then

$$P(A_1 \cup A_2 \cup A_2 \cup \ldots) = \sum_i P[A_i]$$

Probability Rules: Using only the axioms of probability general rules can be derived. In particular, we have for any events A and B

$P[\overline{A}] = 1 - P[A]$

$P[A \cup B] = P[A] + P[B] - P[A \cap B]$

Conditional Probability The chance of event A occurring given that event B occurs, denoted by $P[A|B]$, is called the conditional probability of A given B. Provided $P[B] > 0$

$$P[A|B] = \frac{P[A \cap B]}{P[B]}$$

Independence Events A and B are said to be statistically independent if and only if:

$$P[A \cap B] = P[A]P[B]$$

Note that when A and B are independent $P[A|B] = P[A]$ which says that the event B provides no *conditional* information about the chance of A occurring.

4.3.3 Venn Diagrams

A useful visualization of event relationships is called a *Venn Diagram*. The sample space $\mathcal{S}$ is represented as a rectangle. In Figure 4.1 Disjoint (top left), Overlap (top right), and Triple (bottom) relationships between events A, B, C are shown.

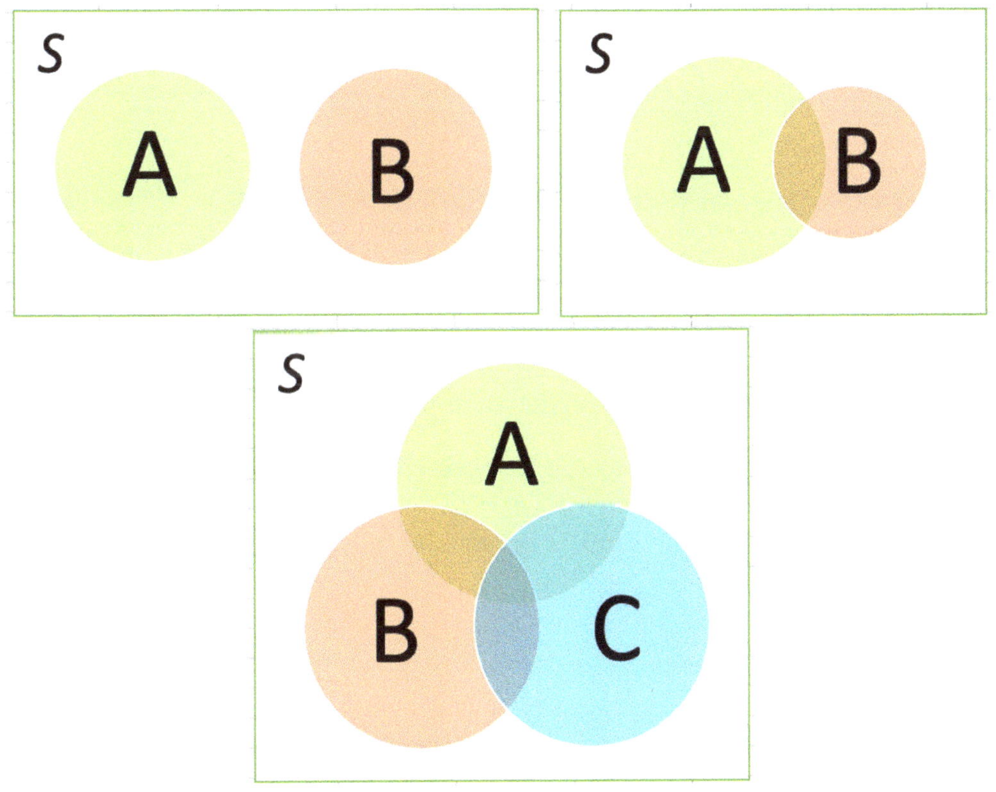

FIGURE 4.1. Venn Diagrams

Key Point: Venn Diagrams show Event relationships — NOT probabilities

4.4 Some Probability Examples

The Coin Toss example makes sense (axiomatically and empirically), but some probability problems can be head-scratchers! Let us look at a few examples

Example 4.3 What is the chance we get 1 Head in 3 tosses of a fair coin?

The possible outcomes from (Toss 1)(Toss 2)(Toss 3) are:

$$HHH, HHT, HTH, THH, \boxed{\text{HTT, THT, TTH}}, TTT$$

Assuming independence and $P[H] = 0.5$ the chance of getting only one H (boxed outcomes) is $3/8 = 0.375$ (37.5%) Certainly a reasonable chance this could occur.

However, extending this approach to $n = 10$ tosses would be **very** *tedious!*

So let's try to generalize the problem:

- For $n = 3$ tosses we had 8 possible outcomes.
- $8 = 2 \times 2 \times 2 = 2^3$ That is, each "Toss" has 2 possible outcomes H or T

  So the **total** number outcomes can be calculated by multiplying together the number outcomes for each individual Toss[1]
- If we **assume** the coin is fair (P[Head] = P[Tail] $= \frac{1}{2}$) then the chance of getting any "one" of the 8 outcome patterns is: $\frac{1}{8} = \frac{1}{2} \times \frac{1}{2} \times \frac{1}{2} = 0.125$
- There are **3 ways** to get one H in $n = 3$ tosses (boxed outcomes)
- Hence, Prob[one H in $n = 3$ tosses] $= 3 \times (0.125) = 0.375$

Example 4.4 Election Results are often predicted *before* all the votes are counted

This may seem presumptive, but let us consider the situation where Candidates A and B are running for an elected position. Suppose $n = 10$ votes have been submitted and after counting seven (7) votes the current result is one of the following scenarios:

1. $A = 5, \; B = 2$
2. $A = 4, \; B = 3$

Since 6 votes are needed to have a clear winner, neither case is decided although the best Candidate B can do is to tie in scenario # 1. Can we predict Candidate A to be the winner in scenario # 1? This depends on the assumptions we make about the 3 uncounted votes.

[1]This assumes the Tosses are *"independent"* — i.e., the outcome of (Toss 1) has no influence on the outcome of (Toss 2)

- *Equally Likely* This is equivalent to Example 4.3 where a fair coin is tossed 3 times. The only way for Candidate B to tie is to get all the remaining votes $\Rightarrow P[TTT] = 1/8$ So Candidate A has an 87.5% $(1 - 1/8)$ chance of winning. A reasonable projection.

- What about scenario # 2? Candidate A needs at least 2 votes to win $= 4/8 = 50\%$ whereas Candidate B needs at least 2 votes $= 4/8 = 50\%$ to tie or win.

  Too close to call.

- *Unequal* If pre-election polls[2] support Candidate A over B — 60% versus 40% say, then Candidate B is less likely to tie in scenario # 1: $P[TTT] = (0.4)^3 = 6.4\%$ Similarly, in scenario # 2, Candidate B chances of a tie or win is reduced to 16.7%

  Conversely, if the pre-election support for Candidate B is $> 50\%$, then the chances of a tie or win for Candidate B will increase in both scenarios.

Example 4.5 Monty Hall Problem[3] Suppose you are on a game show, and you are given the choice of three doors: Behind one door is a car; behind the other two doors are goats. You pick a door, say # 1, and the host, *who* **knows** *what is behind each of the doors,* opens another door, say # 3, which has a goat. He then says to you,

"Do you want to pick door # 2?"

Is it to your advantage to keep or switch your choice?

At first, we might think our chances have increased from 1/3 to 1/2. Initially we choose door # 1 (with a 1 in 3 chance of winning), but now we know the car must be behind our selected door (# 1) or door #2. That is, whether we stay with our original choice or switch, it now appears we have a 50-50 chance of winning.

WRONG !!! We should **switch**

Here is the issue. If you stay with your original choice, then your chances are still 1/3 because *you are ignoring the "conditional" information* — the Host will only open a door that does NOT contain the car. Hence $P[\text{switch}] = 1 - 1/3 = 2/3 = 67\%$

Not convinced? Then let us enumerate all the possible outcomes if you Switch from your original Choice. This is shown in Table 4.1 where we can see that $P[\textbf{win}] = 6/9 = 67\%$

Note that if we switch, there is only one option since we would obviously not switch to the door that was opened. If we are "winning" then Monty has a choice of 2 doors to open. We lose if we switch to the other door.

Example 4.6 Bertrand's Paradox Draw a circle of radius r and insert an equilateral triangle inside with side length ℓ (triangle vertices touch the circle). Randomly drop a straw across the circle and let the "chord" length be c. What is $P[c > \ell]$?

Solutions `https://www.youtube.com/watch?v=uI2FnUmBeeo`

[2]In 2016, 113 of 115 polls projected Clinton would win over Trump in the US Presidential election — Polls can be wrong!
[3]The Monty Hall problem is a probability puzzle named after Monty Hall, original host of the TV show "Lets Make a Deal"

TABLE 4.1. Monty Hall Switch Outcomes

Correct Door	Your Choice	Monty Opens	You Switch	Result
1	1	2 or 3	3 or 2	*lose*
1	2	3	1	**win**
1	3	2	1	**win**
2	1	3	2	**win**
2	2	1 or 3	3 or 1	*lose*
2	3	1	2	**win**
3	1	2	3	**win**
3	2	1	3	**win**
3	3	1 or 2	2 or 1	*lose*

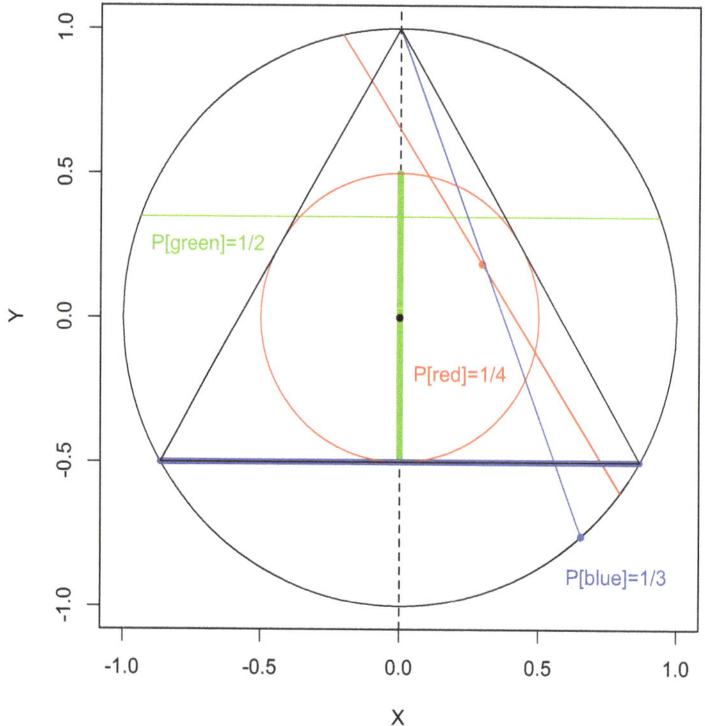

FIGURE 4.2. Bertrand's Paradox Solutions

The paradox arises since dropping a straw "randomly" onto the circle was not well-defined.

If the straw must be dropped horizontally, then the green solution is correct. If one end of the straw must have a fixed starting point then the blue solution is correct. Otherwise the red solution (which uses the center point of the straw) is correct.

4.5 Combinatorics

In Section 4.4 we saw that some (discrete) probabilities were obtained by *counting* procedures. Here we introduce another method of counting called *Combinatorics.*

Definition: Combinatoric The number of ways of selecting x objects from n without regard to the order in which they were selected is given by:

$$\binom{n}{x} = \frac{n!}{x!(n-x)!}$$

where $k! = k(k-1)(k-2)\cdots 3.2.1$ is called k *factorial* and $0!$ is defined as 1
[There is only "1" way to choose "none" $= 0!$ or "all" $= n!$ of the objects from n]

Example 4.7 How many ways can we select 2 people from 10 distinct individuals?

$$\binom{10}{2} = \frac{10!}{2!(10-2)!} = \frac{10.9.(\cancel{8.7.6.5.4.3.2.1})}{2.1(\cancel{8.7.6.5.4.3.2.1})} = \frac{10.9}{2.1} = 45$$

The larger denominator factorial (8! here) will cancel out the tail component of the numerator factorial (10.9.8! here) [It follows that $\binom{n}{n-x} = \binom{n}{x}$ selecting 8 $\equiv$ "not" selecting 2]

Example 4.8 Balls-in-urns problem How many ways can you distribute $n = 5$ red balls (they are indistinguishable) into $r = 3$ distinct urns?

There are 2 solutions: (1) Empty urns NOT allowed, (2) Empty urns are allowed

Solution 1: Empty urns NOT allowed

By brute force we can allocate the remaining 2 red balls in 6 ways. Alternatively, let us visualize the problem. For example, 5 red balls divided into $r = 3$ urns: ○ ○ | ○ ○ | ○

We can put $r - 1 = 2$ dividers | | anywhere between these balls to create $r = 3$ distinct urns with the condition that each divider must occupy a unique position and no divider can be at either end of the balls. There are $n - 1 = 4$ positions for the $r - 1 = 2$ dividers. Hence

$$\binom{n-1}{r-1} = \binom{4}{2} = \frac{4!}{2!(4-2)!} = 6$$

Solution 2: Empty urns allowed

Brute force is possible noting that the $r - 1 = 2$ dividers can be anywhere (including the ends and/or duplicate positions $\Rightarrow$ empty urn) between the $n = 5$ red balls.

Let's pretend $r = 3$ "fictitious" (fake) balls are added so now we have $n + r = 8$ red balls. Then from Solution 1 we have

$$\binom{n+r-1}{r-1} = \frac{7!}{2!(7-2)!} = 21$$

Withdrawing the fictitious balls then allows for some (at most $r - 1$) urns to be empty

For a more detailed exposition of combinatorics, see Ross (2020)

4.6 Random Variables

While the foundations of Probability can be rigorously described using set theory and the probability axioms, the probabilistic nature of a process is more conveniently described in terms of a *random variable.* Here, it is assumed that the outcomes of a process are numeric or can be encoded as such.

Discrete Random Variable

To motivate the notion of a random variable, we consider the coin toss example. By encoding the events T, H numerically as 0, 1 respectively, we can define the coin toss as a "process" X where X is a *variable* taking the value 0 or 1, each with probability 0.5 (hence the qualifier *random*). Notationally, the random nature of this "process" can be fully specified by writing down its *Probability Distribution Function* (pdf):

$$P[X = x] = 0.5 \quad , \qquad x = 0,\, 1$$

Remarks:

- We associate capital X with the "label" assigned to the process and lower case x with an actual value (outcome) of the experiment.
- The notation $P[X = x]$ is often abbreviated to $P_X(x)$ or just $P(x)$

The coin toss is an example of a *discrete* process since X only has positive probability at certain distinct values x Hence X is referred to as a discrete random variable. Other examples of discrete random variables are:

Discrete Uniform: Fair Dice Throw *Equally likely outcomes*
$P(x) = 1/6 \quad , \qquad x = 1,\, 2, \ldots,\, 6$

Geometric: The first occurrence of H in a (fair) coin tossing experiment
$P(x) = 0.5^x \quad , \qquad x = 1,\, 2, \ldots$

Note that in Geometric example, x is the number of tosses required to get the first H event. There is (theoretically) no upper limit to the number of tosses that might be required.

Key Point: A discrete random variable can be defined over the entire real line by setting $P(x) = 0$ at all values x that are not outcomes of the process under consideration.

Properties of $P_X(x)$: *Probability Distribution Function* (pdf)

1. $P_X(x) \geq 0$ for any value x
2. $\sum_x P_X(x) = 1$ The sum over all x's is 1.

Continuous Random Variable

With metric data (eg., temperature, income), events of interest are *intervals* rather than a specific value x. For continuous random variables therefore, probability measures are only associated with continuous intervals on the real line and $P[X = x] = 0$ for any continuous random variable since the single value $\{x\}$ is an interval of zero length.

Since $P[X = x]$ is not meaningful as a probability measure for continuous random variables (since it is an interval $[x, x]$ with zero width), we use a *Probability Density Function* (pdf) $f(x)$ which defines a curve. Thus $f(x)$ is **not** a "probability" but areas under this curve correspond to probabilities of the interval events of interest.

Properties of $f_X(x)$: *Probability Density Function* (pdf)

1. $f_X(x) \geq 0$ for all x
2. The total area under the curve $f_X(x)$ is always 1
3. $P[a < X < b]$ is the area under the curve $f_X(x)$ between $x = a$ and $x = b$.

Note: We will use the notation **pdf** for both discrete and continuous distributions so it is important to recognize what type of process we are dealing with.

Cumulative Distribution Function (cdf)

The cdf of a random variable X is defined by the function $F_X(x) = P_X[X \leq x]$.

With continuous random variables $P[X \leq b] = P[X < b]$ since $P[X = b] = 0$ and thus $P[a < X < b] = F_X(b) - F_X(a)$ For discrete random variables, some care is required in evaluating compound events such as $P[2 < X \leq 4]$ in the dice toss example. Here, the solution is 2/6 since only the values $x = \{3, 4\}$ represent outcomes which have non-zero probability and satisfy the above *interval* inequality.

Numerical approximations in the form of Probability Tables given in Appendix B or from software are needed to evaluate the cdf since $F(x)$ does not have a closed-form expression for many common distributions. The cdf plays an important role in statistics.

1. It provides *critical values* and $P-$values for assessing *significance.*
2. Percentiles are obtained by solving for x_p in $F(x_p) = p$ with $0 < p < 1$
3. The *median* m is of interest in dose-response studies: $F(m) = 0.5$
4. Survival analysis makes extensive use of $1 - F(t) = P[\text{ survives beyond time } t\,]$

Expectation

At the beginning of this chapter we introduced the notion of a parameter such as the population mean μ and standard deviation σ versus a statistic (e.g., $\overline{x}, s$) computed from a sample. It follows that if we assume the ERROR has a specific distribution (pdf), then we can theoretically compute any parameter associated with the ERROR and DATA populations. In much of our inferential statistics, we will be particularly interested in deriving the population mean μ and variance σ^2 of a random variable X.

This can be achieved using the *expectation operator* $E[X]$ which, for a discrete random variable, is defined by

$$\mu \equiv E[X] = \sum_x xP(x)$$

which is a weighted *average* of the values of X. Hence, it represents the *expected value* or mean μ of X. For continuous random variables $E[X] = \int xf(x)dx$ which requires calculus.

The definition of $E[X]$ can also be regarded as an *operator* in the sense that any function $g(X)$ is evaluated as:

$$E[g(X)] = \sum_x g(x)P(x)$$

Note that $g(x)$ only changes the numerical value of x, not the "inherent" probability measure associated with x. In particular, setting $g(X) = (X - \mu)^2$ the (population) variance σ^2 of a discrete random variable X is defined by:

$$\sigma^2 = Var[X] \equiv E[(X - \mu)^2] = \sum_x (x - \mu)^2 P(x)$$

where the standard deviation σ is the positive square root of this expectation. For continuous X we have $Var[X] = \int (x - \mu)^2 f(x)dx$.

4.7 Special Distributions

So far we have only considered some general aspects of probability and random variables. However, the original purpose of this discussion was to develop a methodology which would enable the ERROR component of Equation 4.1 to be investigated in a statistical analysis. In particular, we are interested in the mechanism which generates the randomness which we observe in the DATA. From the above discussion we now know that this randomness can be described in terms of a pdf associated with a random variable. In practice of course, the "true" pdf underlying the process is unknown and so we need to employ some pdf which best reflects the context of the problem. The following list gives some common distributions which cover a range of generic applications.

4.7.1 Discrete: Binomial, Geometric, Poisson

Discrete Random Variables:

Although the context of one process may be quite different from another, the generic encoding may result in the same type of $P(x)$. For example, a coin toss, True/False question, Yes/No opinion, etc. are all processes which can be generically encoded as 0,1 valued random variables. However there are many ways to analyze this discrete data type which gives rise to a variety of pdfs.

Bernoulli: Let X be the random variable associated with a process having only two outcomes, success and failure say, encoded as 1,0 respectively. The chance of a success is denoted by $p = P[X = 1]$. Then X is called a Bernoulli random variable and has pdf:

$$P(x) = p^x(1-p)^{1-x} \quad , \quad x = 0, 1 \qquad \mu = p, \ \sigma^2 = p(1-p)$$

The Bermoulli pdf is the probability distribution for a "single" trial, assuming the success probability $p = P[X = 1]$ does **not** change between trials.

Binomial: If a Bernoulli process is repeated (independently) n times, then the random variable associated with the number of successes $X = x$ in n trials is called a Binomial random variable, denoted by $X \sim \text{Bin}(n, p)$ and has pdf:

$$P(x) = \binom{n}{x} p^x (1-p)^{n-x} \qquad x = 0, 1, \ldots, n \qquad \mu = np, \ \sigma^2 = np(1-p)$$

Geometric: Let X be the number of Bernoulli trials required to obtain the *first* success. Then X has a geometric distribution defined by:

$$P(x) = p(1-p)^{x-1}, \qquad x = 1, 2, \ldots \qquad \mu = \frac{1}{p}, \ \sigma^2 = \frac{(1-p)}{p^2}$$

Hypergeometric: Suppose n items are to be drawn from a hat containing N objects, M of which are labelled as "success" When the selection is done *without* replacement, the chance of getting a success after each draw will change. For this type of process the Hypergeometric distribution is used.

Poisson: When an event is assumed to occur at a certain rate λ over time and the chance of getting more than 1 occurrence of this event within a *small* time interval is negligible, a Poisson distribution can be used to characterize the process. Thus, $X \sim \text{Poi}(\lambda)$ is a discrete random variable which corresponds to the number of times the "event" occurs over a unit time period and has pdf:

$$P(x) = e^{-\lambda}\lambda^x / x! \qquad x = 0, 1, 2, \ldots \qquad \mu = \lambda = \sigma^2$$

4.7.2 Continuous: Uniform, Exponential

Continuous Random Variables: Here we define $f(x)$ and give the mean and variance.

Remember that probabilities will be computed as areas under these curves over continuous intervals.

Uniform: $X \sim U(a, b)$ The continuous equivalent of a discrete process whose outcomes occur with equal probability (eg. a dice toss) is a Uniform process over the interval (a, b). The pdf is:

$$f(x) = \frac{1}{b-a} \quad , \qquad a < x < b \qquad \mu = \frac{a+b}{2}, \ \sigma^2 = \frac{(b-a)^2}{12}$$

Exponential: $X \sim Exp(\lambda)$ The Exponential distribution can be used to model failure-time / survival-time processes (such as radioactive decay) where the exponential decay $\lambda > 0$ is constant. This distribution also satisfies the **memoryless property** wherein knowing that no failure has occurred up to time t does not change the nature of the process. It has no "memory" so to speak, of what has happened prior to time t. The pdf is:

$$f(x) = \lambda e^{-\lambda x} \quad , \qquad x > 0 \qquad \mu = \frac{1}{\lambda}, \ \sigma^2 = \frac{1}{\lambda^2}$$

4.8 The Normal Distribution

This is also called the Gaussian distribution, Bell curve, and is one of the most important distributions used in statistics. Many processes have distributions that can be well approximated by the Normal and most of the estimation and inference theory we shall employ assumes Normality. We therefore need to have a good understanding of the properties of a Normal distribution. The $X \sim N(\mu, \sigma)$ pdf is parametrized by its mean μ and variance σ^2 and defined by the pdf:

$$f(x) = \frac{1}{\sqrt{2\pi\sigma^2}} \exp\left\{\frac{-(x-\mu)^2}{2\sigma^2}\right\} \qquad -\infty < x < \infty$$

Properties:

1. The pdf of X is parametrized by its mean and variance. That is, $E[X] = \mu$ and $Var[X] = \sigma^2$ are the same quantities which appear in the formula for $f(x)$
2. The curve $f(x)$ is bell-shaped and symmetric about $x = \mu$. Thus the population mean, median and mode all occur at $x = \mu$
3. $f(x) > 0$ for all x However, for practical purposes, we should note that:
   - 68.26% of the area under any Normal curve lies in the interval: $\mu \pm \sigma$
   - 95.44% of the area under any Normal curve lies in the interval: $\mu \pm 2\sigma$
   - 99.74% of the area under any Normal curve lies in the interval: $\mu \pm 3\sigma$
4. Any probability measure for X can be evaluated using the *Standard* Normal distribution $Z \sim N(0, 1)$ which has mean 0 and variance 1.

Standardization:

In Appendix B, Tables A.1 and A.2, cumulative probabilities of the standard Normal distribution $Z \sim N(0, 1)$ are tabulated as $F(z) = P[Z \leq z]$. Probabilities for $X \sim N(\mu_x, \sigma_x)$ are evaluated by using the standardization transformation: $Z = (X - \mu_x)/\sigma_x$

Example 4.9 *Find $P[9 < X < 18]$ where $X \sim N(15, 3)$*

Here $\mu_x = 15$ and $\sigma_x = 3$. Thus

$$P[9 < X < 18] = P\left[\frac{9-\mu_x}{\sigma_x} < \frac{X-\mu_x}{\sigma_x} < \frac{18-\mu_x}{\sigma_x}\right] = P[-2 < Z < 1]$$

Using the properties of a cdf we see that $P[-2 < Z < 1] = F(1) - F(-2)$. Thus

$$P[-2 < Z < 1] = F(1) - F(-2) = 0.8413 - 0.0228 = 0.8185$$

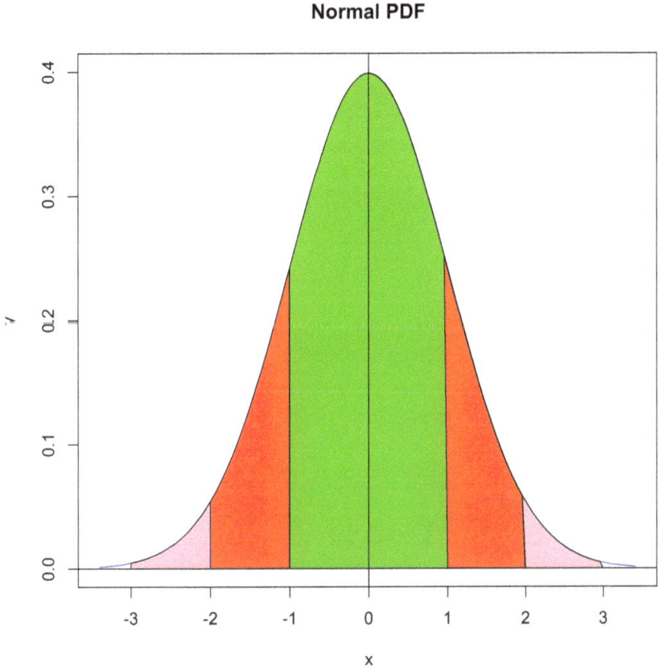

FIGURE 4.3. Standard Normal PDF

4.8.1 Central Limit Theorem (CLT)

Motivation: An investigator wants to monitor components produced from an assembly line in order to assess the "average" daily proportion of the number of defective items, and for non-defective items, what is the "average" lifetime of the component.

It would be impractical to measure *every* component so the assessment would have to based on a (random) sample. However, the aim of the study is to assess the "true" average proportion (ρ_D say) of daily defectives, and similarly, the "true" average lifetime (μ_L say) of the non-defective components. This seems like an intractable situation.

Key Point: The CLT (defined below) will enable us to *infer* something *meaningful* about the (unknown) population mean [a proportion is the mean of binary (0,1) counts]

Sound too good to be true? We have two very distinct processes — the number of defectives is count data so the distribution is discrete, whereas the distribution of lifetimes will be positive and continuous. Since these population pdfs are unknown we cannot compute the quantity $\mu = E[X]$. Nevertheless, the CLT statement above does hold true and illustrates how important the CLT result is for statistical inference.

Theorem: Central Limit Theorem (CLT)
Let $X_1, X_2, \ldots, X_n$ be a *random sample* from a process X having mean μ and variance σ^2 and $\overline{X}$ be the mean of the random sample. Then, as n increases, the following random variable converges in distribution to a standard Normal random variable.

$$Z = \frac{\overline{X} - \mu}{\sigma/\sqrt{n}} \rightarrow N(0,1)$$

Remarks:

1. Knowledge of the underlying pdf of the process X is not required.
2. In general, the approximation becomes very good for $n > 30$
   [The result is exact for any n if the process $X \sim N(\mu, \sigma)$]
3. By *unstandardizing* it follows that $\overline{X} \rightarrow N(\mu, \frac{\sigma}{\sqrt{n}})$

Note: Remark 3 shows that the *expected* mean and standard deviation of $\overline{X}$ satisfy:
$E[\overline{X}] = \mu$ and $\sqrt{Var[\overline{X}]} = \sigma/\sqrt{n}$ However, to avoid confusion with the usual notion of "standard deviation" we will refer to $\sigma/\sqrt{n}$ as the **standard error** (SE) of $\overline{X}$

The importance of the first remark is that knowledge of the *actual* distribution of the population is **not** needed. The distribution can be discrete or continuous, skewed, bimodal, or almost[4] any type of distribution which is unlikely to be known in practice. The second remark shows that the CLT can be realistically applied in practice since the approximation is sufficiently good for sample sizes that are cost-effective to obtain.

Problem: The third remark would appear to make the CLT result completely "irrelevant" since it requires we know the true population mean μ and standard deviation σ which is clearly not the case in practice when we only have a random sample from the population. Catch-22 ? First, let's see if $\overline{X}$ does appear to converge to a Normal distribution.

Example 4.10 Consider the Exponential distribution of Figure 4.4(a).

This is strongly right skewed and clearly not a Normal distribution. In plot (b) we have taken the average of $n = 2$ randomly selected values from plot (a). This was done many times and plot (b) is the resulting histogram of all the averages $\overline{X}_{(n=2)}$ from the samples of size $n = 2$. Similarly, plots (c) and (d) are the resulting histograms of all the averages $\overline{X}_{(n)}$ from samples of size $n = 10$ and $n = 25$ respectively, randomly selected from plot (a).

Plot (d) includes the fitted Normal curve (green line) which is very close to the smoothed density curve (red line). Thus, $\overline{X}_n$ does appear to converge to a Normal distribution as n increases and we will take this as evidence that the CLT result holds.

[4]Technically, the mean and variance need to be finite

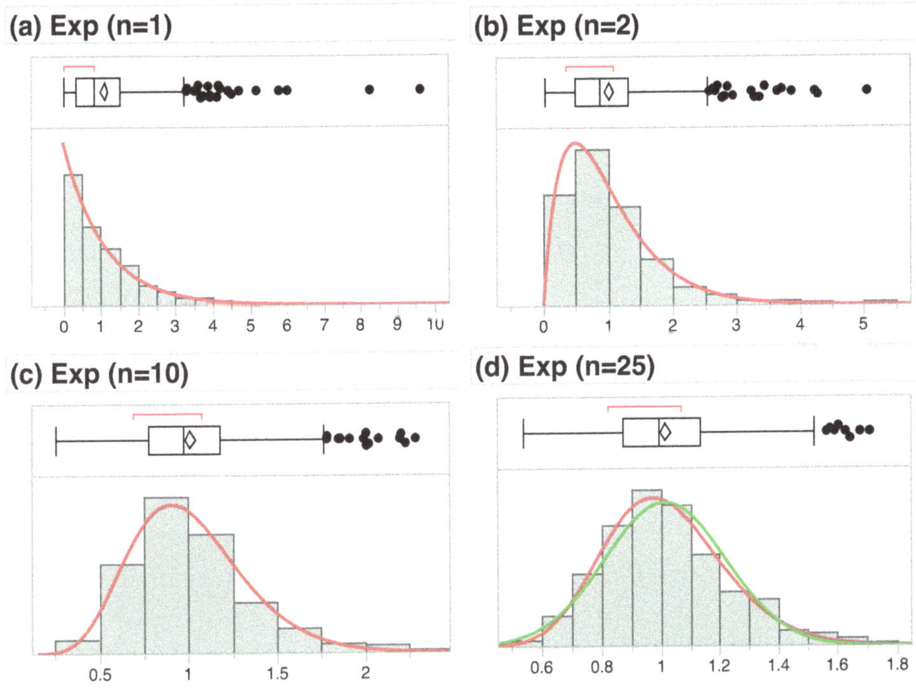

FIGURE 4.4. CLT for $\overline{X}_n$ from Exponential

4.8.2 Normal Approximation to the Binomial

When $n > 20$ computing the exact value of $P(x)$ when $X \sim Bin(n,p)$ requires statistical software. For example, let $X \sim Bin(n = 20, p = 0.3)$ and we want to compute $P[X < 5]$

$P[X < 5] = P[X = 0] + P[X = 1] + \cdots + P[X = 4] = \binom{20}{0}p^0q^{20} + \binom{20}{1}p^1q^{19} + \cdots + \binom{20}{4}p^4q^{16}$

where $q = 1 - p = 0.7$

Key Point: The CLT can help us simplify the problem.

Provided both np and nq are ≥ 5 we can approximate the Binomial distribution X by

$$Y \sim N(\mu = np, \sigma = \sqrt{npq})$$

Hence, with $np = 6$ and $nq = 14$ (both ≥ 5) and $npq = 4.2$ and using the **continuity correction** $P[X = x] = P[x - 0.5 < Y < x + 0.5]$ we have

$$P[X < 5] \approx P[Y < 4.5] = P[Z < (4.5 - 6)/\sqrt{0.42} = -0.73] = 0.2327 \quad \text{from Table A.1}$$

From R: `pbinom(4,20,0.3)` $= PX \leq 4] = 0.2375$ so the Normal approximation is quite close (less than 0.005 difference!)

4.9 Computing Probabilities and Quantiles

The Probability Tables in Appendix B are useful, but somewhat tedious to use. Ideally, we would like to answer the following scenarios using software:

Let the pdf of X be specified.

- Given the *Quantile* "x" what is the *Probability* $p = P[X = x]$ (or $P[X < x]$ etc.)
- Given the *Probability* "p" $= P[X = x]$ what is the *Quantile* "x"

So what are our options?

Applets Bognar (2021) provides a free applet for computing a variety of distributions.

Textbook Some publishers bundle propriety software (such as StatCrunch) with their Text

R This can be downloaded for free and provides a comprehensive suite of statistical methods and pdf's

JMP Can compute probabilities and quantiles. Menu-driven.

Key Point: The pdf parameters and quantile or probability needs to be entered.

Example 4.11 Binomial probability $P[X \leq 4]$ using Bognar's Applet. We entered $n = 20$, $p = 0.3$, and $x = 4$ into the boxes as shown. The popup menu allows us to select the probability we derived in section 4.8.2

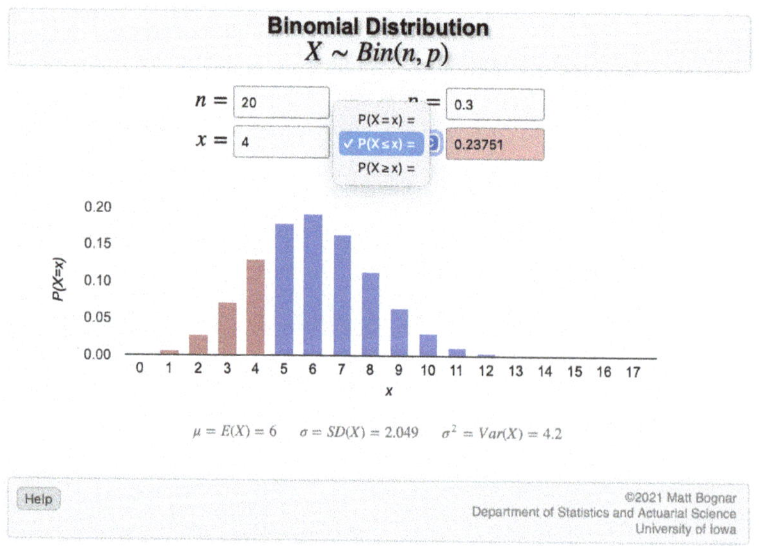

FIGURE 4.5. Binomial Applet

R software: R has a a wide range of pdf distributions. For a probability we use:

`pdistname(q,parameters)` (e.g., `pbinom(4,20,0.3)` $= PX \leq 4] = 0.2375$)

For a quantile we use:

`qdistname(p,parameters)` (e.g., `qnorm(p=0.975)` $= 1.959964$ or 1.96 rounded)

The caveat is that you need to know what acronym R uses for the `"distname"` We will use R sparingly throughout this text.

JMP Notes: JMP is menu-driven so certain tasks (such as computing probabilities) can take several steps. We consider the simpler example of $X \sim Bin(n = 4, p = 03)$ with x as the column of quantiles that want to compute the Binomial probabilities

First we create the JMP data table and create the formula for **prob** as shown in Figure 4.6 Note that JMP puts $P[X = 5]$ as missing since can not have $x = 5$ when $n = 4$.

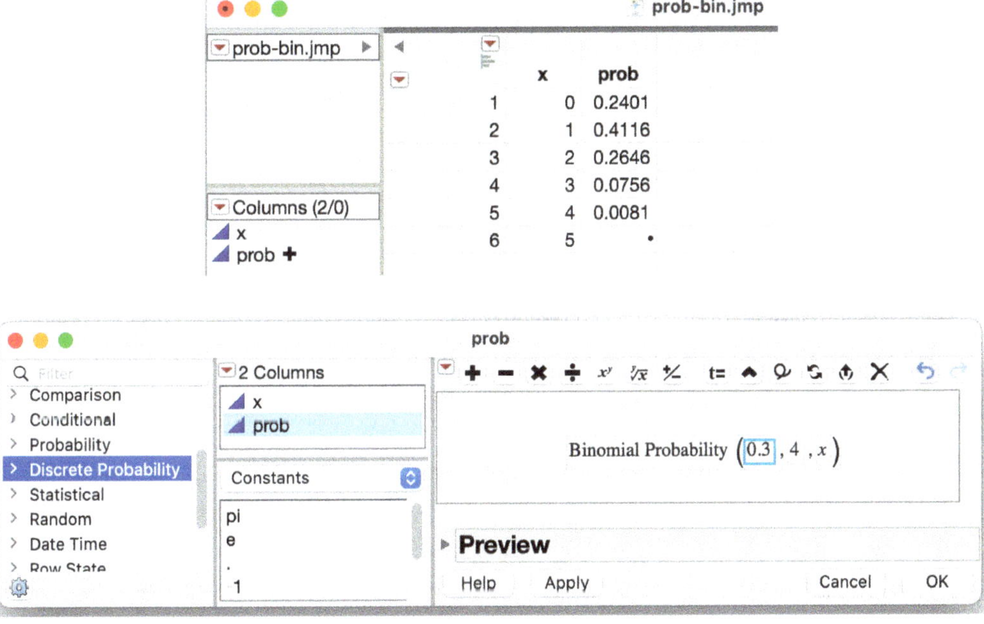

FIGURE 4.6. Binomial JMP DataTable and Formula

Chapter 4 Exercises

4.1 The following sports injuries occurred to players on a team during the season:
Totals: Arm = 5 , Leg = 7 , Neck = 5

Within these Totals, the number of players who had multiple injuries were:
Arm & Leg = 4, Arm & Neck = 3, Leg & Neck = 2, and Arm & Leg & Neck = 2

Construct the Venn diagram for this data. How many players were injured during the season?

4.2 Let A, B and C be any nontrivial events. Give expressions for the following:

a) only A occurs. [ANSWER: $A \cap \overline{B} \cap \overline{C}$ or $A \cap \overline{(B \cup C)}$]

b) both A and C occur, but B. does not occur.

c) at least one of the events occur.

d) at least two of the events occur.

e) all three occur.

f) none of the events occur.

g) at most one of them occurs.

h) at most two of them occurs.

i) exactly two of them occur.

j) at most three of them occur.

4.3 Given that $P[A] = 0.5$ and $P[A \cup B] = 0.6$, find $P[B]$ if

a) A and B are *independent.*

b) $P[A\,|\,B] = 0.5$

c) With reference to part (b) above, is it possible for C be an event which is disjoint from A, but such that $P[C\,|\,B] > 0.5$? Explain.

4.4 Refer to the **PULSE** data. Assume the 39 individuals constitute a random sample.

a) What is the probability that ALL the observed pulse rates are even?

b) Hence, what is your *practical* conclusion?

c) How could the assumption of independence be violated? Explain.

4.5 A person has $20,000 to invest in four possible opportunities. Each opportunity will accept investments in $1,000 units, but only after the required minimum investment has been made. The minimums are $2, $2, $3 and $4 thousand, respectively. How many investment strategies are possible if:

a) An investment must be made in each opportunity?

b) Investments are made in *at least* three opportunities?

4.6 Discuss the basic conclusion of the "Birthday Problem" [Web search]

a) What assumptions are typically made?

b) Hence, what is the chance that from four (4) different people, at least one pair have the same **mm/dd** birthday?

4.7 Memoryless property
Suppose the customer waiting time X at a supermarket cashier follows the Exponential distribution: $X \sim Exp(\lambda)$ From calculus we can show that $P[X > x] = e^{-\lambda x}$

a) Show that $P[X > s + t\,|X > t\,] = P[X > s\,]$

b) Use the cashier scenario to express this result in words.

NOTE: The Geometric distribution also satisfies the memoryless property

4.8 Let $X \sim N(\mu = 100, \sigma = 5)$ Use Tables A.1, A.2 to find the following probabilities
NOTE. **Tail** probabilities "outside" of the range $(-3.5, 3.5)$ can be regarded as "0"

a) $P[X = \mu\,]$

b) $P[X < 90\,]$

c) $P[\,86.5 < X < 108.4\,]$

d) $P[\,90 < X < 120\,]$

4.9 Let $X \sim N(\mu = 100, \sigma = 5)$ Use Tables A.1, A.2 to find the quantile x
[HINT: If an "exact" value does not exist, make an *appropriate* estimate.]

a) $P[X < x\,] = 0.5$

b) $P[X > x\,] = 0.05$ [HINT: Convert the inequality]

c) The 90th percentile of X [i.e., $P[X < x\,] = 0.9$]

4.10 A random number generator, X say, is supposed to produce random numbers that are uniformly distributed on the interval from 0 to 1. That is, $X \sim U(0, 1)$. "IF" this is true, the numbers generated should come from a Uniform population with $\mu = 0.5$ and $\sigma = 0.2887$ (see `runif` graph in Figure 3.8).
Suppose $n = 100$ generated numbers from X resulted in a sample mean $\overline{x} = 0.4365$

a) Verify that $\sigma = 0.2887$ (rounded to 4 decimal places)

b) i) What is the approximate distribution of $\overline{X}$ when $n = 100$?
ii) How reliable (good) is this approximation? Explain.

c) Hence compute $P[\overline{X} < 0.4365]$

d) What does this suggest about the *quality* of the random number generator X

Part II

Bivariate Analysis

5

Correlation and SLR

Is X related to Y ? Let's find out.

Suppose you want to buy a house. You know many factors play a role in buying a House, but one certainly stands out: bigger = more expensive. Hence, we expect X = square footage is related to Y = selling price. Similarly, we could analyze whether there is a Gender ($X = M, F$) effect on SAT performance ($Y = score$), or which Hospital ($X = A, B$) has the best survival rate after major surgery ($Y = survived, died$), or does Exercise ($X = time\ spent\ exercising$) improve your health ($Y = yes, no$). The basic methods for analyzing these scenarios are:

Scenario	X	Y	Method
House	Quantitative	Quantitative	Correlation, Regression
SAT	Categorical	Quantitative	t–tests, ANOVA
Hospital	Categorical	Categorical	Contingency Table
Exercise	Quantitative	Categorical	Logistic Regression

Key Point: Individual observations in these examples consist of two "paired" measurements (x_i, y_i) on the same subject i

That is, the *two* measurements (x_i, y_i) represent **one** observation for Subject = "$\boldsymbol{i}$" since we need to use **both** the X and Y variables simultaneously in order to assess whether a relationship exists based on the bivariate methods above.

Caution: The context of the study will be important in our interpretation of the results from a bivariate analysis. A statistical association between X and Y does **not** imply *causality*[1]. Similarly, the lack of association does not necessarily imply X and Y are unrelated.

In this chapter we focus on **Correlation** and simple linear **Regression** (SLR) methods.

[1]See "silly correlations" link in section 4.2

5.1 Scatterplots

In the House example, X = square footage versus Y = selling price are individually measured for the **same** house. That is, the square footage (x_i) of your house is specifically tied (paired) to the selling price (y_i) of **your** house.

Regression Dataset Let X and Y both be quantitative variables where the observed values x_i and y_i are directly *paired* (same subject). Then the dataset $\{(x_i, y_i)\}$ $i = 1, 2, \ldots, n$ is called a **regression** dataset.

Scatterplot The visualization diagnostic for a regression dataset is the Scatterplot. Here, we plot each point (x_i, y_i) as co-ordinates on the standard Cartesian X, Y axes graph.

Example 5.1 House Prices The dataset and scatterplot for Square footage versus selling Price (in \$1000s) of 12 houses in a local neighborhood are shown in Figure 5.1.

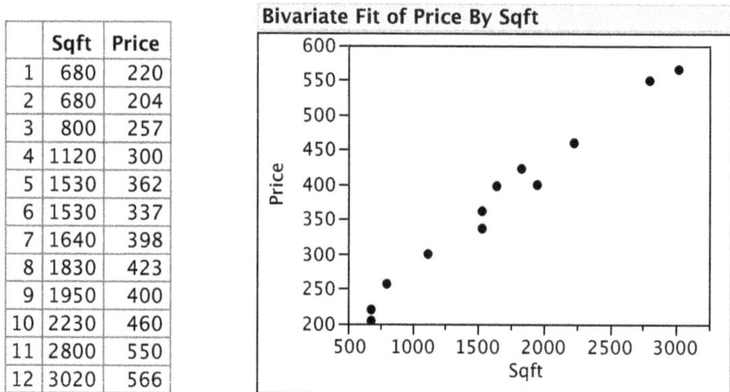

	Sqft	Price
1	680	220
2	680	204
3	800	257
4	1120	300
5	1530	362
6	1530	337
7	1640	398
8	1830	423
9	1950	400
10	2230	460
11	2800	550
12	3020	566

FIGURE 5.1. House Example

The scatterplot shows what we would expect: *as the size of the houses get larger, the selling price increases.* However, we can also see there are exceptions to this general trend. Houses with the same square footage sold for different prices and in one case, a larger house (`Sqft` = 1950) sold for less than a smaller house (`Sqft`=1830).

Key Point: For a regression dataset we focus on the general **trend**

Other factors such as number of bathrooms, bedrooms, and lot size affect House prices, so the exceptions noted above would be expected within the range of the general upward trend.

Some more examples of scatterplots are shown in Figure 5.2. Plots (a), (b), (d) and (e) each show an obvious **pattern** and we need to develop some terminology to describe the underlying features of interest in these scatterplots. Equivalently, we also need to be able to describe situations such as shown in Plots (c) and (f). That is, we should be able to *verbally* communicate the relevant features so that a knowledgable colleague will have a good visual understanding of the display without actually seeing the scatterplot.

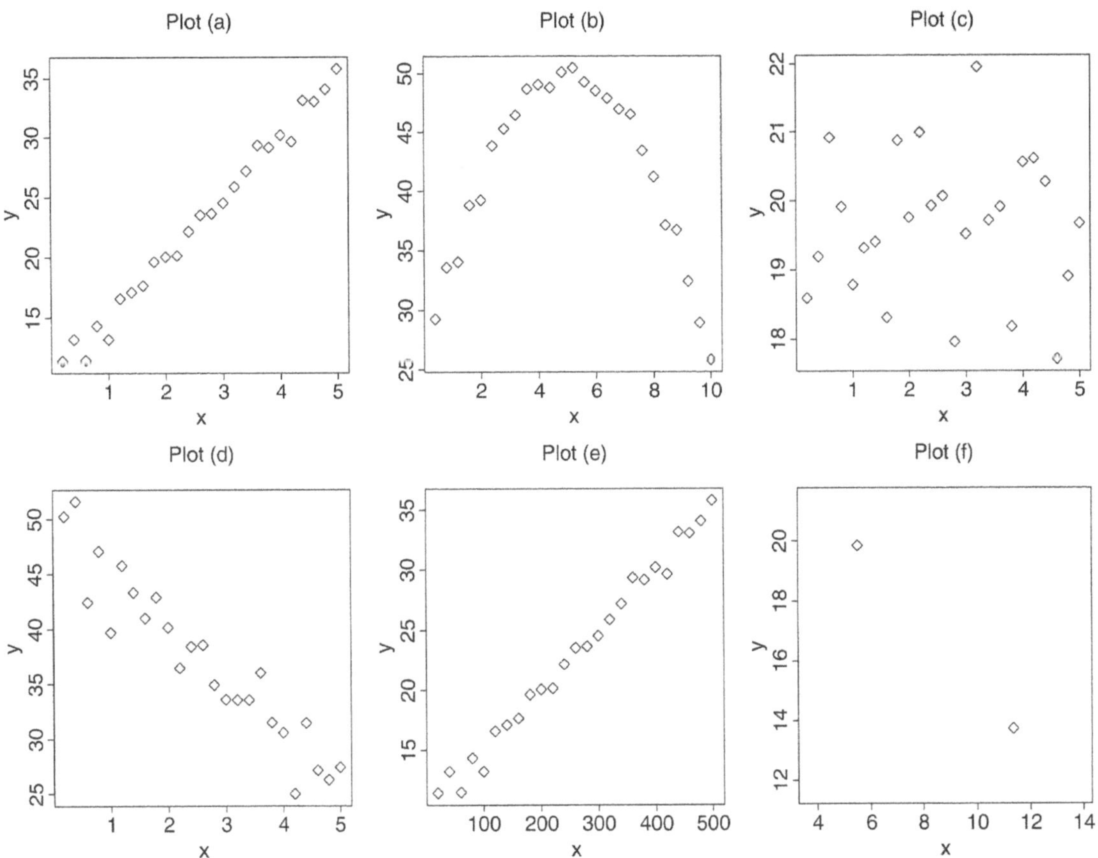

FIGURE 5.2. Scatterplot Examples

- **Trend** This describes the underlying pattern of the scatterplot.

Linear A straight line would *summarize* the upward trend in Plots (a) and (e). The points in Plot (d) are not as **tight** to the downward trend, but a straight line would still be a good description of the downward trend. Thus, we use the term **Linear** when a straight line provides a reasonable summary of this type of pattern in a scatterplot.

IMORTANT: We **NEVER** use a *vertical* straight line to summarize a trend since this assumes the X values do not change. This does not represent a regression dataset.

Nonlinear Plot (b) shows curvature. Whenever there is obvious curvature, we use the term **Nonlinear** to describe the trend. To be more specific, the following terminology presented in Figure 5.3 can be used to describe certain nonlinear trends as appropriate.

NOTES: The *hook* of the J-curve distinguishes it from the Exponential curves. Also the Logarithmic curve can have negative $y-$values whereas the Square Root cannot. A Polynomial curve (such as the Cubic) is more of a generic term for a nonlinear trend that is "anything else."

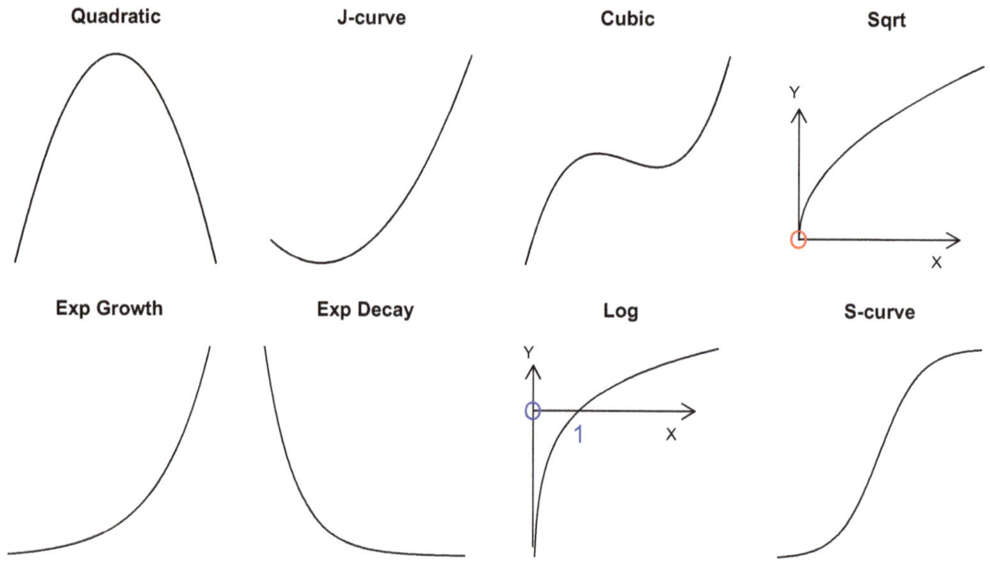

FIGURE 5.3. Nonlinear Curves

- **Association** If a scatterplot exhibits a **linear** trend, we use the term **Positive** association when the trend is upward and **Negative** association when the trend is downward.

  Key Point: The **direction** of the *linear* association is always relative to: "as X increases."

  Hence if Y increases as X increases $\Rightarrow$ **Positive** association. Plots (a) and (e)
  If Y *decreases* as X increases $\Rightarrow$ **Negative** association. Plot (d)

  Plot (b) clearly shows an association (initially Positive, then Negative), but this will be confusing to our colleague. We will be better served by describing this scatterplot as having a quadratic association. Similarly for other nonlinear trends.

  Plot (c) shows no discernible pattern (neither linear or nonlinear). In this case we can describe this scatterplot as having **No association**

- **Strength** Now we want to describe how *close* the points in the scatterplot lie to the underlying trend we identified. Plots (a), (b) and (e) are *tighter* than Plot (d). Thus we use the terminology **strong** to **weak** to convey the notion of strength of association.

  Key Point: To evaluate the strength of association we must ignore the X and Y scales

  That is, strength of association is a *relative* measure which means Plots (a) and (e) look identical [*They are. The X scale was simply $\times$ by 100 in Plot (e)*]

  And finally to Plot (f). Since a downward straight line would fit both points exactly, we have **perfect** (negative) association. That is, maximum strength of association.

  To **quantify** the strength and direction of association between X and Y we need a *numerical* summary statistic which we present in the next section.

5.2 Correlation

The scatterplot provides us with a *visual* diagnostic for evaluating the features of a regression dataset that we discussed above. Now we develop a numerical measure that quantifies the strength and direction of the **linear** association between X and Y in a regression dataset. This measure is called the *correlation coefficient* and is denoted by the letter r (the term "coefficient" is often omitted and we simply say r = correlation).

Key Point: r is a measure of the **linear** association between X and Y

Hence, r would NOT be an appropriate measure for Plot (b) which is clearly nonlinear (quadratic) with a strong association. Indeed, the value of r computed for Plots (b) and (c) will be close to 0 even though the scatterplots are very different. As always, visual diagnostics should be inspected as numerical summary statistics can be misleading.

Correlation Coefficient Let $\{(x_i, y_i)\}_{i=1}^n$ be a regression dataset. The correlation between X and Y is given by:

$$r = \frac{1}{n-1}\sum_{i=1}^{n}\left\{\left(\frac{x_i - \overline{x}}{s_x}\right)\left(\frac{y_i - \overline{y}}{s_y}\right)\right\}$$

We would obviously use JMP to compute this quantity in practice !

Properties of Correlation

1. $r_{xy} = r_{yx}$ The correlation measure makes no distinction between X and Y so r is the same if we had plotted `Price` on the x-axis and `Sqft` on the y-axis in Figure 5.1

2. r has no units; it is just a number This follows from the fact fact that both X and Y are *standardized.* Hence Plots (a) and (e) have exactly the same value of r since changing the scale of X and/or Y makes no difference.

3. $r > 0 \Rightarrow$ Positive association In the House example we saw that as `Sqft` increased, so did the `Price`. Hence $r > 0$ for Plots (a) and (e)

   Similarly, $r < 0 \Rightarrow$ Negative association as in Plot (d)

   Key Point: $r = 0$ only implies no **linear** association

   This is true of Plot (c), but is misleading for Plot (b) which has a strong nonlinear (quadratic) association.

4. $-1 \leq r \leq 1$ This can be shown to hold for any regression dataset

   Thus, values of r near 0 indicate a very weak linear relationship — Plot (c), whereas values closer to ± 1 indicate a stronger linear relation — Plots (a), (e) and Plot (d)

   At the extremes, $r = -1$ or 1, ALL the (x_i, y_i) points must lie *exactly* on a downward or upward sloping straight line. Hence $r = -1$ for Plot (f)

From the following examples shown in Figure 5.4, it can be seen that there will **always** be some variation about a linear trend no matter how close r gets to ± 1 ($r = 0.999$ scatterplot). This variation will be large for $-0.7 < r < 0.7$ and as r gets closer to 0, it may be hard to perceive or justify that a linear trend even exists ($r = -0.2$ scatterplot).

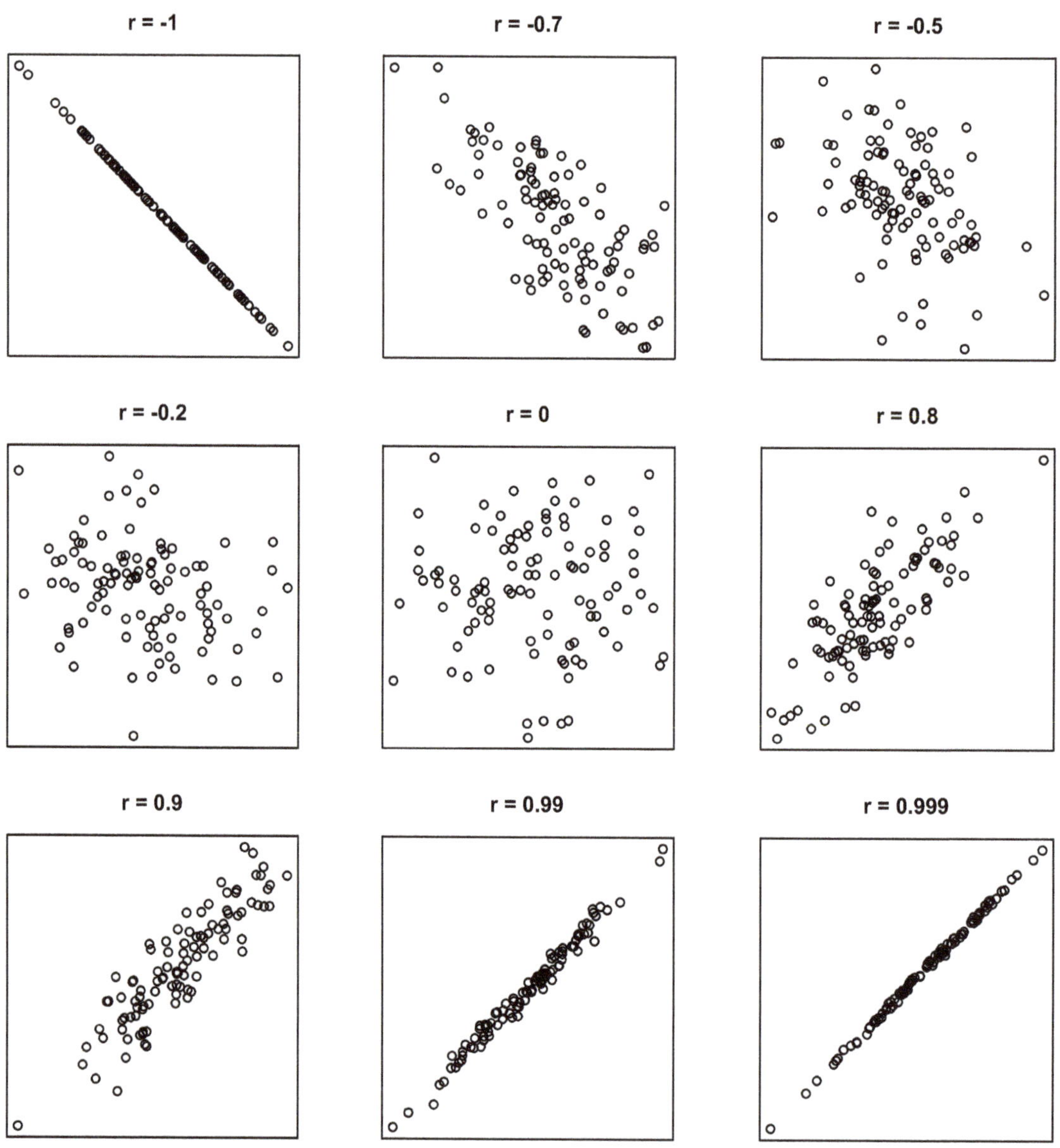

FIGURE 5.4. Correlation Examples

Caution: Be aware that correlation can be misused !

Examples of silly correlations: http://www.tylervigen.com/spurious-correlations

5.3 Simple Linear Regression (SLR)

In the House example, $r = 0.99$ suggesting a strong linear association and so a straight line would summarize the upward trend very well. A straight line would also be a reasonable description of the underlying downward trend in Plot (d) even though there will be more variation around this line. Hence, when a scatterplot of a regression dataset shows a **linear** trend, we could *summarize* the linear relationship by actually drawing a **straight line** on the scatterplot. This procedure is called **Simple Linear Regression** (SLR for short) and while it is closely related to correlation, SLR differs in one important respect:

Key Point: *For SLR the distinction between X and Y is critical.*

5.3.1 Straight Line

The equation for a straight line is $Y = a+bX$ where a, b are constants (called *parameters*) that uniquely define a straight line. Specifically,

$a =$ *intercept* the value of Y when $\boldsymbol{X} = \mathbf{0}$, and
$b =$ *slope* the change in Y for a **unit increase** in X

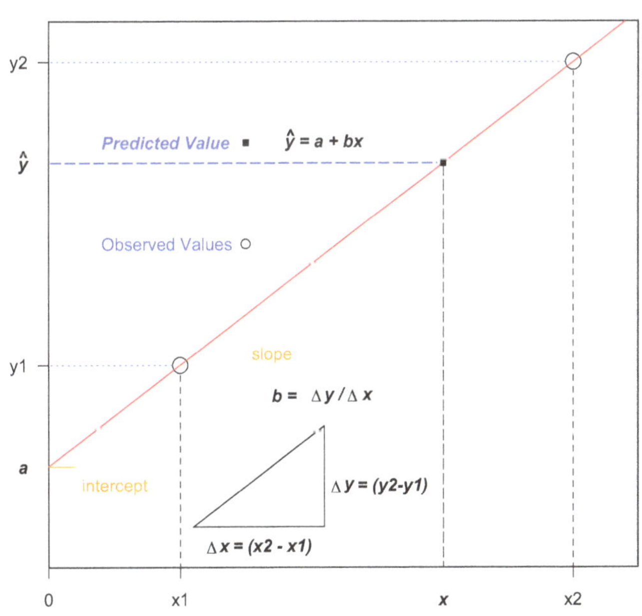

FIGURE 5.5. Straight Line Equation

Example 5.2 Let $x1 = 1$, $x2 = 4$ and $y1 = 3.5$, $y2 = 5$ in Figure 5.5
Then $\boldsymbol{b} = (5 - 3.5)/(4 - 1) = 0.5$ and $\boldsymbol{a} = y1 - b * x1 = 3.5 - 0.5 * 1 = 3$
We get the same result for a using $x2$, $y2$ — or *any* point on the straight line.
For example, the *Predicted Value* at $\boldsymbol{x} = 3$ is: $\hat{\boldsymbol{y}} = 3 + (0.5) * 3 = 4.5$

5.3.2 Regression Line

Fitting a straight line to **summarize** a linear trend exhibited by a **regression** dataset such as the House example in Figure 5.1 presents a more difficult problem. Any line of the form $\hat{y} = a + bx$ cannot pass through *all* the points in Figure 5.1 , so how do we fit the **best** straight line to summarize this type of data? Figure 5.6 shows the SLR line that JMP (somehow) fitted resulting in the *prediction* equation:

$$\texttt{Price} = 124.3 + 0.15 * \texttt{Sqft} \tag{5.1}$$

where the **intercept** $a = 124.3$ and **slope** $b = 0.15$ (rounding both parameter **estimates** from the Linear Fit report in Figure 5.6)

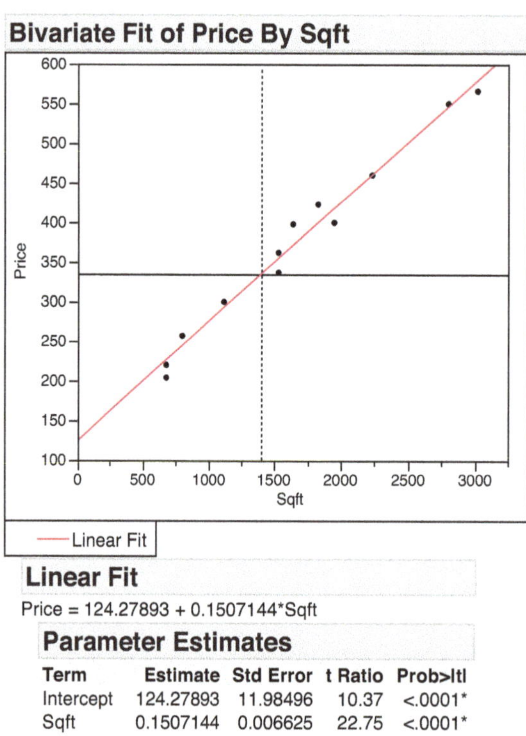

Linear Fit

Price = 124.27893 + 0.1507144*Sqft

Parameter Estimates

Term	Estimate	Std Error	t Ratio	Prob>|t|
Intercept	124.27893	11.98496	10.37	<.0001*
Sqft	0.1507144	0.006625	22.75	<.0001*

FIGURE 5.6. House Example: Fitted SLR

The SLR fitted **regression line** looks reasonable and allows us to predict the selling `Price` of a house for any given `Sqft` value, such as 1400 shown by the crosshairs in Figure 5.6

$$124.3 + 0.15*(\texttt{Sqft} = 1400 \text{ square feet}) = 334.3 = \text{predicted } \texttt{Price} \text{ (in \$1000s)}$$

Key Point: The regression line $\hat{y} = a + bx$ can be used to **predict** Y given $X = x$

Note that we are using the notation "$\hat{\boldsymbol{y}}$" (called *y-hat*) to distinguish the SLR fitted line (red line) from the actual observed values of y (black dots). Now we interpret the SLR fit.

5.3.3 Interpretation of the SLR

The equation $\hat{y} = a + bx$ used for the regression line has the same two parameters a, b that were needed to uniquely define a straight line in Figure 5.5 Recall that

$a =$ *intercept* is the value of y when $x = 0$, and
$b =$ *slope* is the change in y for a **unit increase** in x

From the fitted SLR line in Figure 5.6, we can deduce the following:

- The intercept $a = 124.3$ represents the predicted `Price` in \$1000s for a *virtual* House with `Sqft` $= 0$ (!?) — which obviously doesn't make sense in the context of this dataset.

- While the value of the intercept a may not have a meaningful statistical interpretation, it is mathematically necessary to uniquely define a line. The key point to remember is that the intercept is *always* defined at $x = 0$ — it is NOT (in general) where the regression line crosses the y-axis.

- Figure 5.6 was stretched out so that `Sqft`=0 was included in the display. Fitting the regression line (equation 5.1) to Figure 5.1 would cross the y-axis at $y = 199.3$. This is the predicted `Price` at `Sqft` $= 500$ which is **not** the intercept value.

- The slope b is therefore the pertinent parameter of interest. In Figure 5.6, $b = 0.1507$ which is **positive** and so for this dataset, Houses cost \$150.70 per square foot on average. That is, each additional square foot will *increase* the predicted `Price` by \$150.70 = 0.1507*\$1000 since `Price` is measured in units of \$1000s.

- A straight line implies a **constant** rate of change in y with respect to x. Hence we could also state that the predicted `Price` increases by \$150,700 per 1000 square feet.

5.3.4 SLR Notation

The advantage of a regression line is that it is a simple **model** and easy to interpret once the intercept and slope have been determined. The problem is that our observed regression dataset will deviate from this model because `Price` is not completely determined by `Sqft`. That is, most of the points do not lie on the straight line $y = a + bx$ so this notation is somewhat misleading from a statistical perspective since $y_i \neq a + bx_i$ in general. To remedy this situation we adopted the standard *hat* notation so that $\hat{y} = a + bx$ represents any point on the fitted regression line. Henceforth, we refer to a predicted value of Y at $X = x$ as "**y-hat**" Some additional notation is defined below and illustrated in Figure 5.7

SLR Model:	$y_i = a + bx_i + e_i$	
Fitted SLR:	$\hat{y} = a + bx$	equation of the regression line
Predicted value:	$\hat{y}_o = a + bx_o$	predicted value of Y at a particular $X = x_o$
Fitted value:	$\hat{y}_i = a + bx_i$	predicted value of Y at **observed** points $X = x_i$
Residual:	$e_i = y_i - \hat{y}_i$	difference between Observed and Fitted Value

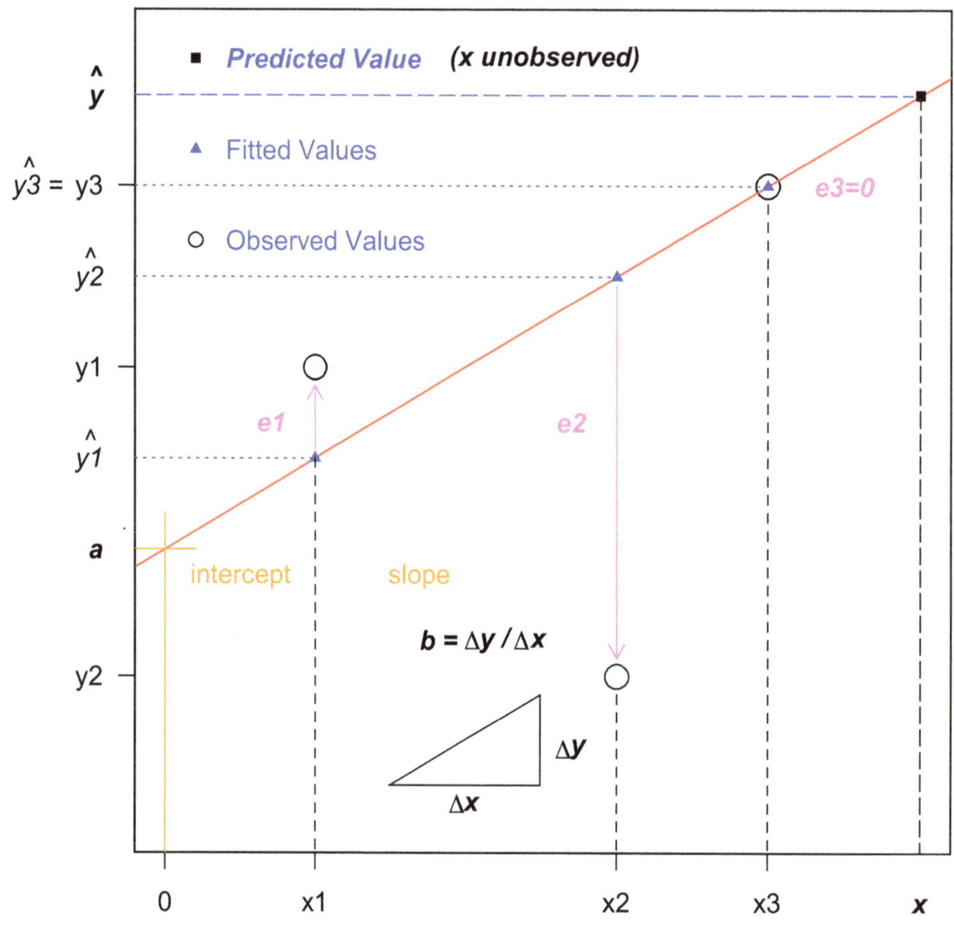

FIGURE 5.7. SLR Notation

Of course, none of the above quantities can be computed until $\boldsymbol{a}, \boldsymbol{b}$ have been determined. We shall do this shortly, but let us first check that fitting a SLR is appropriate. For example, it would not make sense to fit an SLR to Plot (b) in Figure 5.2 since it shows a nonlinear (quadratic) trend. Figure 5.8 shows more scatterplot examples that we consider below.

Plot 1 is an example of a **growth** curve which is often used to model population growth. Although we fitted a straight line (which does reflect the positive association), it is clearly inappropriate in terms of predicting Y — systematically underestimating Y at small and large values of X, and overestimating in between.

Plot 2 illustrates the **funnel effect** or *heteroskedasticity* where the variability of Y increases as X gets larger. While the underlying trend is essentially linear, our ability to predict Y becomes less precise as X increases. Hence a SLR model would **not** provide an adequate description of this process since it doesn't account for the important heteroskedastic feature

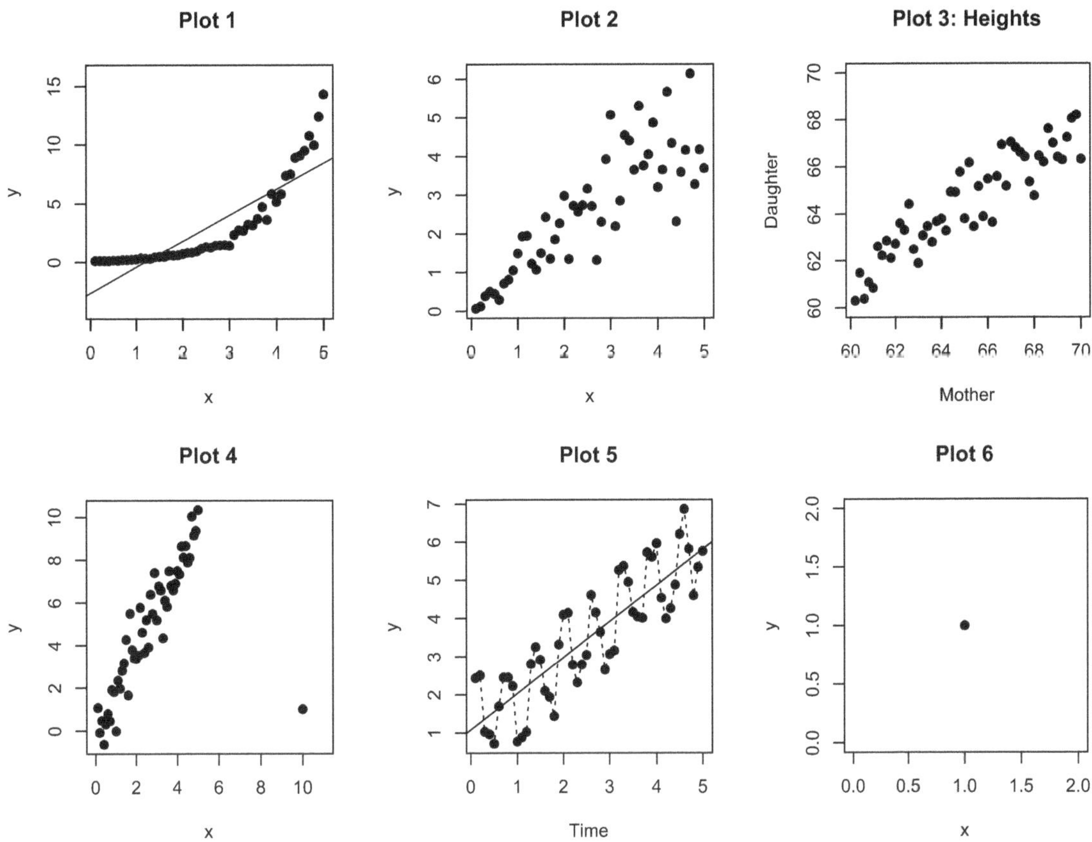

FIGURE 5.8. More SLR Examples

Plot 3 represents the Mother – Daughter equivalent of the Father – Son heights study conducted by Sir Francis Galton who coined the phrase **regression to the mean**. In a study of 1000+ Father and Son heights, he found that the Son's height on average *regressed* towards the mean. That is, the Son of a Father who was taller than average would also be tall, but not as tall as his Father (on average). Similarly, the Son of a Father who was shorter than average would also be short, but not as short as his Father (on average). The same result was shown with Mother – Daughter heights. The SLR model is appropriate here.

Plot 4 shows an ***X* outlier** ($X = 10$, $Y = 1$) *[note that the $Y = 1$ value is **not** an outlier with respect to the range of Y]* This could be a typo ($X = 1.0$ would make this point consistent with the linear trend) or indicative of a *threshold* process where the Y values rapidly decrease after some X between 5 and 10. (Such as the discharge of a capacitor.) Without the outlier, a SLR model would be appropriate.

Plot 5 would seem like a simple upward linear trend. However, when X is **known** to be recorded sequentially or represents a `Time` index, a **Timeplot** of the time series can reveal cyclic patterns. See Figure 3.14. Here, the SLR model does **not** capture the cyclic behavior.

Plot 6 emphasizes that we can not fit a SLR with **insufficient** information. The same would apply to Plot (f) in Figure 5.2 in the (statistical) sense that we should **not** expect a "new" observation to lie *exactly* on a line fitted to the two points currently shown.

5.4 Least Squares Estimation (LSE)

We have a regression dataset $\{(x_i\,,\,y_i)\}_{i=1}^{n}$ (with $n > 2$). Now we want to fit the regression line $\hat{y} = a + bx$ which requires computing estimates for the parameters a, b. Figure 5.6 shows the result using JMP for the Housing example which seems quite reasonable (the fitted line follows the upward trend closely and looks to be placed in the *middle* of the points). However, Figure 5.8 shows that fitting a SLR model to a regression dataset may not always be appropriate. So we need certain assumptions to be satisfied before fitting a SLR.

5.4.1 SLR Assumptions

Linear trend: We assume the regression dataset follows the **linear** model

$$y_i = \boxed{a + b\,x_i} + e_i \tag{5.2}$$

where a, b are the intercept and slope parameters (to be determined). As indicated specifically in Figure 5.7 some of the observed points $(x_i\,,y_i)$ will "deviate" from the *fitted* SLR regression (red line). This deviation is measured by the **residual** (violet font and arrows)

$$e_i = y_i - \hat{y}_i \tag{5.3}$$

where $\hat{y}_i = a + b\,x_i$ is the **fitted** value (boxed term in equation 5.2 and blue triangles in Figure 5.7). If the *linear* assumption is satisfied then the residuals e_i (equation 5.3) should not show any obvious pattern with respect to x_i

Equal variation: We also assume that the observed points exhibit **equal variation** about the fitted regression line. That is, the y_i values should lie within a "band" of constant width which follows the linear trend as depicted in Figure 5.9, Plot (a)

We also added a *density ellipse* (green shaded region) in Plot (a) to illustrate that equal variation is an **overall** exploratory visualization assessment with regard to our *band* of constant width (blue dotted lines). Thus, Plot (a) is not perfect, but it certainly appears to satisfy the equal variation assumption, whereas Plot (b) clearly does not.

Key Point: Do NOT use a **microscope** to assess these SLR assumptions

5.4.2 Fitting the SLR

The House example in Figure 5.6 suggests the deviations from the fitted regression line (equation 5.1) should *balance-out.* That is, the residuals e_1, $e_2, \ldots,$ e_n satisfy a **balance**[2]

[2]This is analogous to $\overline{x}$ being the balance point of a univariate dataset which satisfies $\sum(x_i - \overline{x}) = 0$

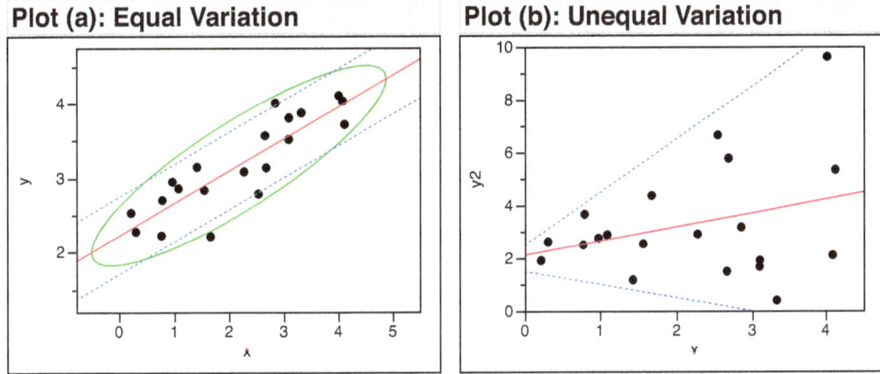

FIGURE 5.9. (a) Equal Variation (b) Unequal Variation

criterion such as $\sum e_i = 0$ which hopefully forces the fitted line toward the "middle" of the linear trend. Unfortunately, this is *not* sufficient as shown by the following example.

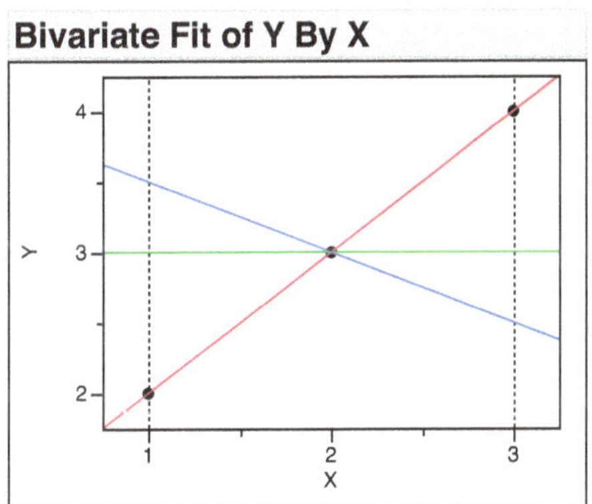

FIGURE 5.10. Known SLR

This regression dataset consists of the three points (1,2), (2,3), (3.4) which all lie exactly on the red line in Figure 5.10 and so the red line is clearly the correct SLR regression line that we would want JMP to compute. But ... there is a problem.

Problem: Both the green and blue lines *also* satisfy the balance criterion $\sum e_i = 0$

Example 5.3 From Figure 5.10 $e_1 = e_2 = e_3 = 0$; $e_1 = 1, e_2 = 0, e_3 = -1$ and $e_1 = 1.5, e_2 = 0, e_3 = -1.5$ all of which satisfy $\sum e_i = 0$ [as will any line through (2,3)]

However, using the *magnitude* of the residuals $\sum |e_i|$, or the **squared** residuals $\sum e_i^2$ from each line, immediately identifies the red line as the **best** fit. For technical reasons, we will estimate the parameters a, b by minimizing the sum-of-squared errors (residuals) SSE $= \sum e_i^2$ From calculus, the Least Squares Estimates (LSE) for a, b that minimize SSE are:

$$a = \overline{y} - b\overline{x} \qquad \text{and} \qquad b = r\frac{s_y}{s_x} \tag{5.4}$$

Key Point The LS regression line is $\hat{y} = a + bx$ with a, b as above

Remark The reason why the distinction between X and Y is critical to SLR is that the residuals e_i are only measured **vertically**. That is, $\hat{y} = a + bx \not\Leftrightarrow \hat{x} = -a/b + y/b$

5.4.3 Properties of LS Regression

1. The distinction between X and Y is critical in SLR.

   The Remark above addresses this important distinction from a mathematical perspective. In practice, this distinction may not always be clear (e.g., Height vs Weight), or it may simply be what we regard as more pertinent: House `Price` versus `Sqft`. The point is that we need to make the choice as to what regression variable is the regarded as the response Y

2. Correlation and Slope are associated

   $b = r(s_y/s_x)$ so $b > 0 \Leftrightarrow r > 0$, $b < 0 \Leftrightarrow r < 0$ and $b = 0 \Leftrightarrow r = 0$

   since s_x, s_y are both positive quantities (for nontrivial regression datasets). This means Positive association $\Rightarrow$ positive slope and correlation; Negative association $\Rightarrow$ negative slope and correlation; No (linear) association $\Rightarrow r = 0$ and the best SLR fit is a horizontal line ($b = 0$)

3. The LS fitted SLR regression line always passes through the point $(\overline{x}, \overline{y})$

   A SLR regression line is like a seesaw with the pivot point at $(\overline{x}, \overline{y})$ Useful for assessing influential points.

4. The LS fitted SLR regression line is balanced: $\sum e_i = 0$

   The LSE a, b that minimize SSE also satisfy the balance criterion $\Rightarrow \overline{e} = 0$. That is, the LS residuals are centered around 0

5. RSquare $= r^2$ for the LS fitted SLR regression line

   See discussion in the next section

5.5 Residual Diagnostics

In the SLR case, the scatterplot of Y versus X is our primary visual diagnostic for evaluating whether (or not) a linear model is appropriate. Curvature, outliers, or other deviations from a linear trend can be readily seen. However, the scatterplot only provides *pre-fit* information. It is also important to assess the *post-fit* result since our **model** may need modification.

Model Paradigm: `DATA` $\rightarrow$ `MODEL` $\rightarrow$ `ERROR`

In the SLR model we have $y_i = a + b\,x_i + \boxed{e_i}$ where the LSE for a, b in Equation 5.4 provides the `MODEL` component ($\hat{y}_i = a + b\,x_i$). Now we focus on the `ERROR` component: $e_i = y_i - \hat{y}_i$ that we refer to as **residuals**

5.5.1 Residual versus Fitted Plot

A visual diagnostic for assessing `MODEL` **adequacy** is a plot of e_i versus $\hat{y}_i$ with the residuals going on the y-axis. This is called a **Residual versus Fitted Plot** For the LS fitted SLR regression line, the following properties hold.

- The plot of e_i versus x_i will be visually identical for a SLR since $\hat{y}_i$ is just a linear transformation of x_i
- The **average** residual $\bar{e} = 0$ (since the LS fitted SLR regression line is balanced)

Examples of residual versus X plots are shown in Figure 5.11.

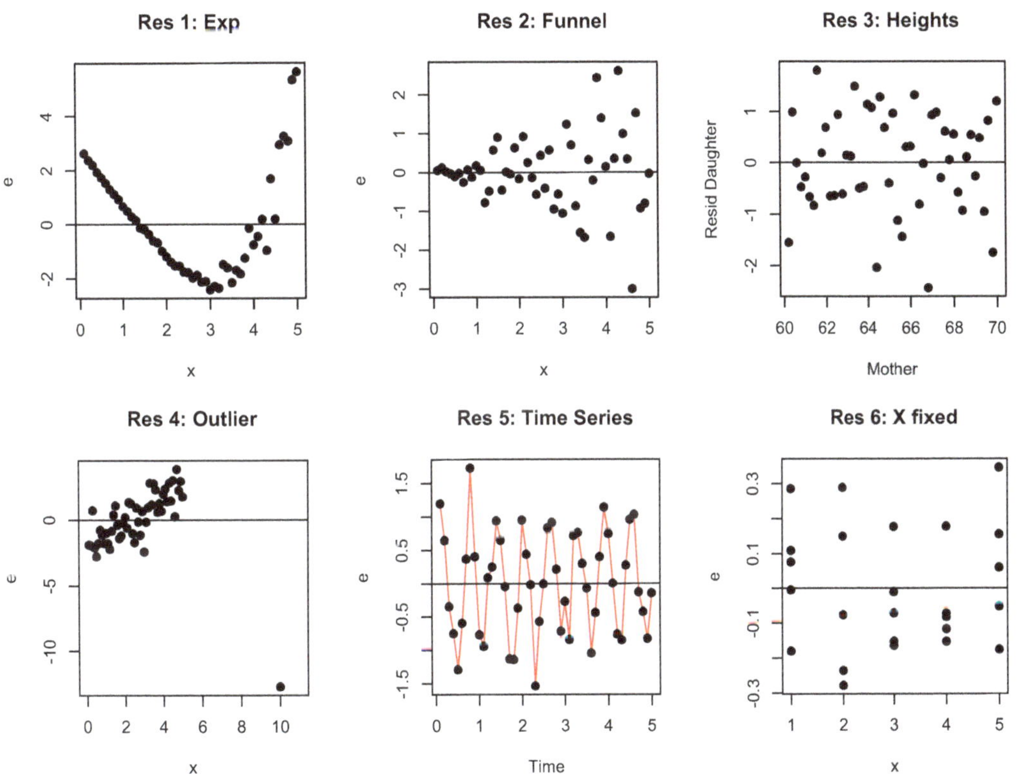

FIGURE 5.11. Residual versus X Examples

The first five residual plots correspond to fitting a SLR to **Plot 1** through **Plot 5** of Figure 5.8 Plots **Res 1, 2, 4, 5** show that the Residuals e_i depend on X indicating that the SLR model is **not** adequate. In contrast, SLR model appears appropriate for the Mother-Daughter dataset since the **Res 3** plot shows no obvious pattern and the residuals exhibit equal variation over the range of Mother heights. The **Res 6** plot shows a *horizontal* structure in that X only takes specific values. However, this is not relevant for assessing SLR adequacy which is concerned with *vertical* patterns of the residuals. Thus, we see that the residuals appear randomly scattered at each X "group" and the variation of the residuals is approximately the same across the X groups. Hence a SLR would be suitable for this regression dataset.

5.5.2 RSquare

Correlation is a unitless measure of the direction and strength of the *linear* association between X and Y in a regression dataset. **RSquare** is a unitless measure of the **quality** of fit of the **model** used to explain the variation of the response Y For example, RSquare would be an appropriate measure of the strength of association of a quadratic *model* fitted to Plot (b) whereas correlation is misleading since the relationship is not linear. Thus,

$$\text{RSquare} = \frac{\text{Variation of } Y \text{explained by the Model}}{\text{Total variation of } Y}$$

It can be shown that $0 \leq \text{RSquare} \leq 1$ where RSquare $= 1$ implies a perfect fit and values less than 1 show decreasing quality of fit. RSquare is often reported as a percentage.

Key Point In the SLR case $\text{RSquare} = r^2$

Thus, the correlation can be determined as $r = \text{sign}(b)\sqrt{\text{RSquare}}$ where $\text{sign}(b) = \pm 1$ (or 0) according to whether the *slope* b of the fitted SLR is positive or negative (or zero).

5.6 JMP Notes for SLR

We now discuss the JMP steps needed to produce the output for the SLR regression line fitted to the House dataset shown in Figure 5.6. First, we create Figure 5.1 Scatterplot:

1. Use the **Analyze** → **Fit Y by X** platform
2. Put `Price` in the Y, Response box and `Sqft` in the X, Factor box.
3. Click **OK** to get the Scatterplot of Figure 5.1

5.6.1 Scatterplot Options

To add results to the Scatterplot, we use the popup menu (red triangle) beside the title Bivariate Fit of Price By Sqft as shown in Figure 5.12 Some useful options are:

Fit Line Fits the LS regression line

Fit Polynomial Submenu allows fitting a quadratic or higher order polynomial
[Not relevant for the House dataset, but fitting a quadratic to Plot (b) in Figure 5.2 would make sense.]

Density Ellipse Any submenu option adds the Correlation report. (See Figure 5.13)

Group By ... Used to add multiple lines for each level of a *Group* variable
[Not relevant for the House dataset. See Example 5.4 later.]

Script Selecting **Redo Analysis** gives a new Output window with the same analysis reports, but reflecting changes made to the dataset (e.g., a point was excluded)

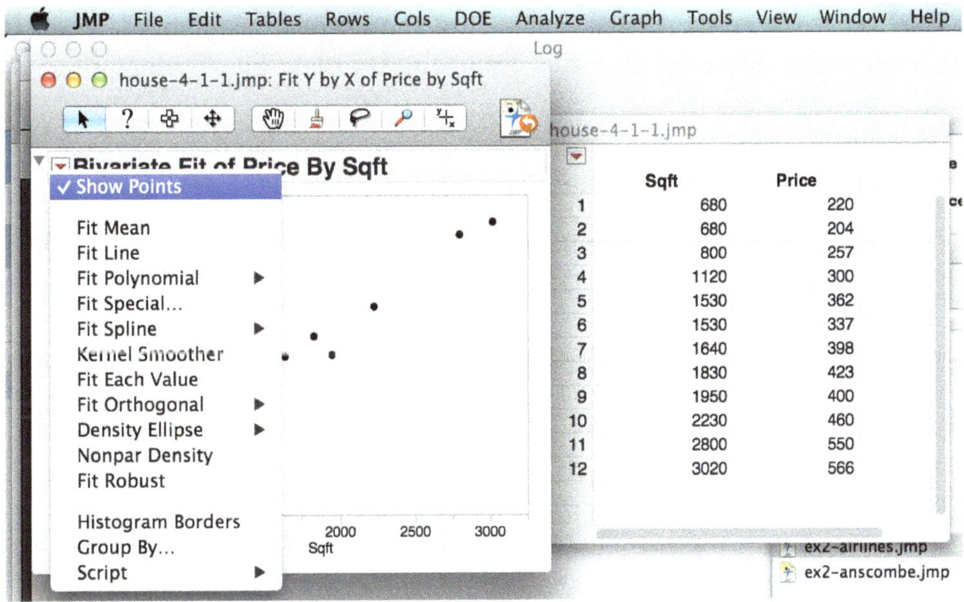

FIGURE 5.12. Bivariate Popup Menu

5.6.2 SLR and Correlation Results

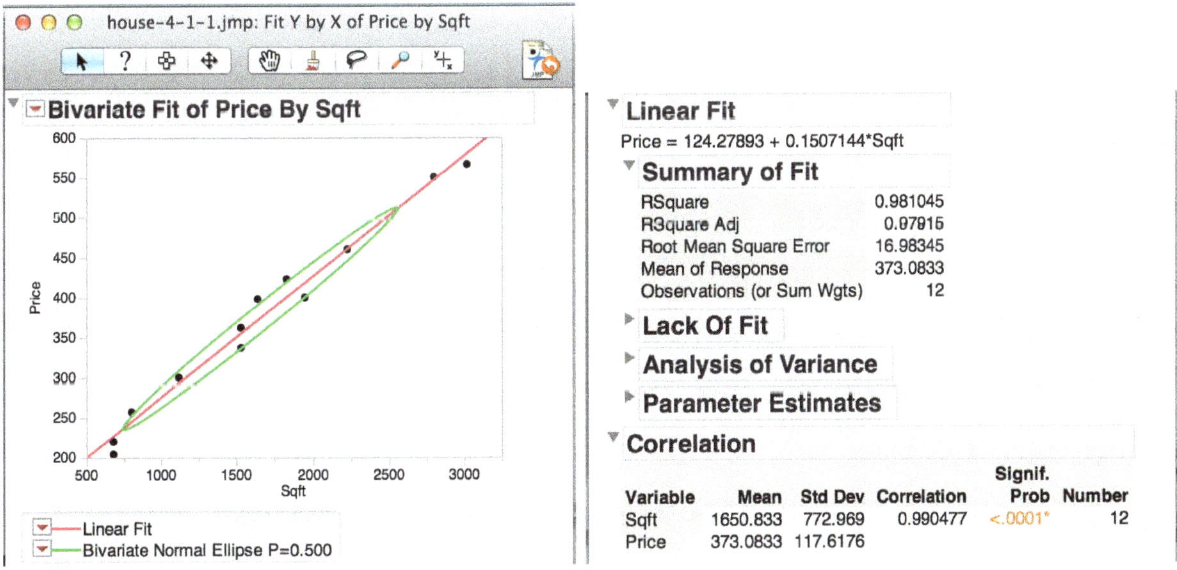

FIGURE 5.13. SLR Regression Line and Correlation Reports

Figure 5.13 shows the amended scatterplot and reports generated after fitting the SLR regression line and correlation to the House data[3] Note that any report section that is closed (right pointing triangle ▸ beside Lack Of Fit) will **not** appear on a printed version of the

[3]The Linear Fit report on the right of Figure 5.13 would actually appear under the Bivariate Fit of Price By Sqft plot.

JMP output (including the title). Hence, make sure the Correlation report is open (triangle down ▾) *before* printing or saving to a file. From the Bivariate Fit ... popup menu, do

1. Select **Fit Line**
   (red line appears on the scatterplot and Linear Fit report generated)

2. Go back to the Bivariate Fit ... popup menu Select **Density Ellipse → 0.50**
   (green ellipse appears on the scatterplot and Correlation report generated)

The SLR regression equation $\texttt{Price} = 124.27893 + 0.1507144 * \texttt{Sqft}$ is shown in Figure 5.13 under the Linear Fit report. The correlation $r = 0.990477$ appears in the Correlation report which very close to 1 as indicated by the narrow ellipse on the scatterplot. It follows that `Sqft` is a very good predictor of `Price` for this House dataset. The Summary of Fit report provides the following quantities:

RSquare As can be verified, RSquare $= 0.981045 = r^2 = 0.990477^2$

RSquare Adj (To be discussed in a later chapter)

Root Mean Square Error This is the standard deviation of the residuals $s_e = 16.98345$ which is computed by $s_e = \sqrt{\text{SSE}/(n-2)}$ [The $(n-2)$ denominator is due to the fact that we need to estimate the 2 parameters a, b in order to compute the residuals]

Mean of Response $\overline{y} = 373.0833$ which the average House `Price`

Observations $n = 12$ is the number of observations in the House dataset.

Example 5.4 Group By ... option. Consider the JMP designed experiment dataset shown in Figure 2.2 where we generate a scatterplot of Y = `PostTrt` vs X = `PreTrt` Using the **Group By ...** option allows us to select `AntiDep` as the grouping variable and fit 3 separate SLR regression lines together on the scatterplot as shown in Figure 5.14 To create this plot the JMP steps are as follows

1. **Analyze → Fit Y by X**

2. Put Y = `PostTrt` and X = `PreTrt` in the dialog window. Click **OK**

3. From the Bivariate Fit ... popup menu Select **Group By ...**

4. Select `AntiDep` from the popup dialog list of variables

5. Now select **Bivariate Fit ... → Fit Line** *voila !*

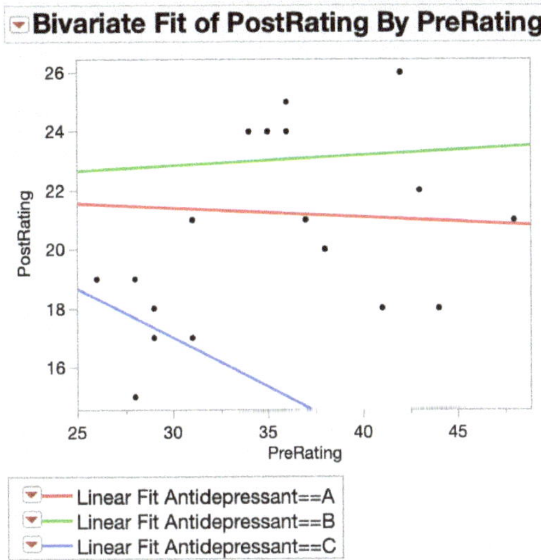

FIGURE 5.14. SLR Regression Lines using Group By ...

5.6.3 Post-fit Options

On the bottom left side of Figure 5.13, additional popup menus (red triangles) are now available under the plot such as beside Linear Fit To add any of these options simply select from the menu options shown in Figure 5.15 For example, we should always select Plot Residuals to check for outliers or nonrandom patterns.

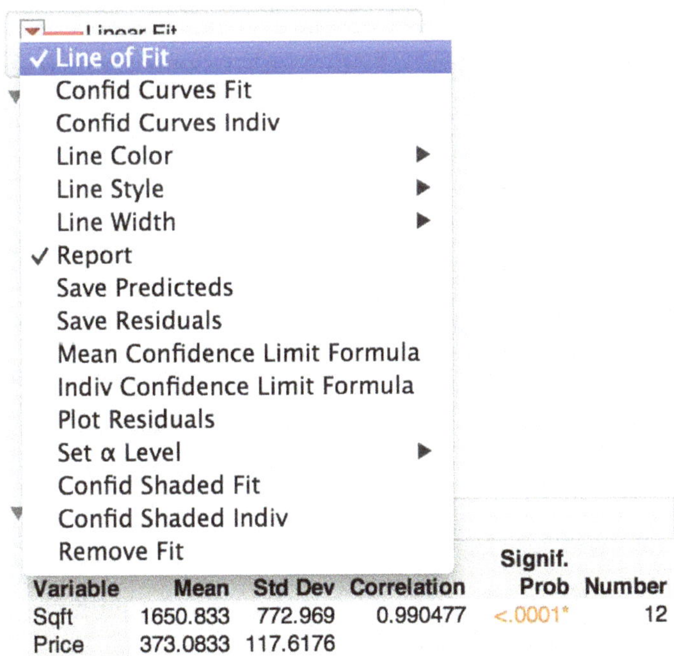

FIGURE 5.15. Linear Fit Options

5.7 Transformations to Linearity

The underlying trend of the growth curve shown in **Plot 1** of Figure 5.8 can be described by the exponential function: $y = A\exp\{bx\}$ (where $\exp\{u\} \equiv e^u$) which is nonlinear. However, the "log-transformation" of this equation gives $\log y = \log A + bx$ which is the straight line $Y = a + bx$ where $Y = \log y$ and $a = \log A$

Applying a transformation to a regression dataset can therefore simplify our analysis in certain situations by **linearizing** the relationship between y and x.

Power Transformation

The class of *power transformations* is defined[4] by

$$y(\lambda) = \begin{cases} \frac{y^\lambda - 1}{\lambda} & ,\lambda \neq 0 \\ \log y & ,\lambda = 0 \end{cases}$$

This is often simplified to $y(\lambda) = \text{sign}(\lambda)y^\lambda$ $,\lambda \neq 0$ and $y(0) = \log y$ so the **identity** transformation ($\lambda = 1$) gives $y(\lambda) = y$. The division by λ or use of $\text{sign}(\lambda)$ is to maintain the same *order* when $\lambda < 0$. For example, if $y_i = 1, 2, 4$ then $y_i^{-1} = 1, \frac{1}{2}, \frac{1}{4}$, but $-y_i^{-1}$ has the same ordering as y_i. So when is a power transformation useful and how do we choose λ ?

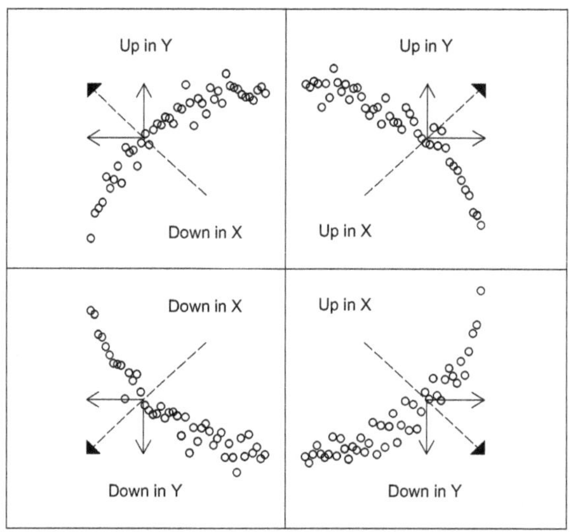

FIGURE 5.16. Power Transformation Schematic

Figure 5.16 illustrates four situations where power transformations are useful for linearizing or *straightening out* the trend. Here, **Up** refers to $\lambda > 1$, and **Down** refers to $\lambda < 1$. To see why this works, consider $y = 4, 16, 100$. Then $y(\lambda = \frac{1}{2}) = \sqrt{y} = 2, 4, 10$ so the square root transformation decreases larger values of y more rapidly than smaller values: $4 \rightarrow 2$ versus $100 \rightarrow 10$. Conversely, $\lambda > 1$ will increase larger values more than smaller values.

[4]It can be shown that $\lim_{\lambda \to 0}(y^\lambda - 1)/\lambda = \log y$

TABLE 5.1. Ladder of Powers

Power	$y(\lambda)$	Name	Context
$\lambda < -1$		—	Not commonly used
-1	$-1/y$	Reciprocal	Inverse e.g., hyperbola
$-1/2$	$-1/\sqrt{y}$	Reciprocal square root	Inverse Area or Counts
$-1/3$	$-1/\sqrt[3]{y}$	Reciprocal cube root	Inverse Volume e.g., mpg
0	$\log y$	Log	Population growth
1/3	$\sqrt[3]{y}$	Cube root	Volume
1/2	$\sqrt{y}$	Square root	Area or Counts
1	y	Identity	No change
2	y^2	Square	Increase
$\lambda > 2$		—	Not commonly used

Now for the choice of λ. First, let's consider the general situation wherein we wish to transform a regression dataset $\{(x_i, y_i)\}$ so that a SLR model $y_i(\lambda_y) = a + bx_i(\lambda_x) + e_i$ would be appropriate. That is, we will entertain power transforming y_i or x_i (or both) using Figure 5.16 as guidance as well as the context of the regression dataset.

In practice, we usually transform y alone, or x alone, and select a power transformation from the **Ladder of Powers** in Table 5.1 Thus, this is a heuristic *trial-and-error* procedure which is intended only to provide a linear simplification for predictive purposes. The choice of λ may be subjective, so we need to be pragmatic. A power transformation can be useful for exploratory purposes, or contextually meaningful (whereas $\sqrt[7]{y}$ is **not** meaningful !)

Power Transformation Notes

1. Negative values are obviously a problem for the Square root and Log transformations. Similarly, the Reciprocal and Log transformations are not defined if $y_i = 0$.

   In some cases adding a constant c such that $y_i + c > 0$ for all observations can remedy this situation ($c = 1$ for Log is often used for zero y_i).

   Note Negative values in nonlinear relationships may have a different underlying dynamic from positive responses. A power transformation may not be appropriate here.

2. Forcing a power transformation onto a regression dataset is (statistically) counterproductive. That is, $\sqrt[5]{y}$ versus $x^{1.23}$ may well linearize the dataset, but how do you justify this choice in the **context** of the dataset. Similarly, using powers $\lambda < -1$ or $\lambda > 2$ can artificially induce linearity simply due to the magnitudes associated with these re-scalings. For example, if $y_i \in 1$ to 100, then y_i^5 ranges from 1 to 10 billion !

3. Be pragmatic. Power transformations are only useful in certain situations.

Chapter 5 Exercises

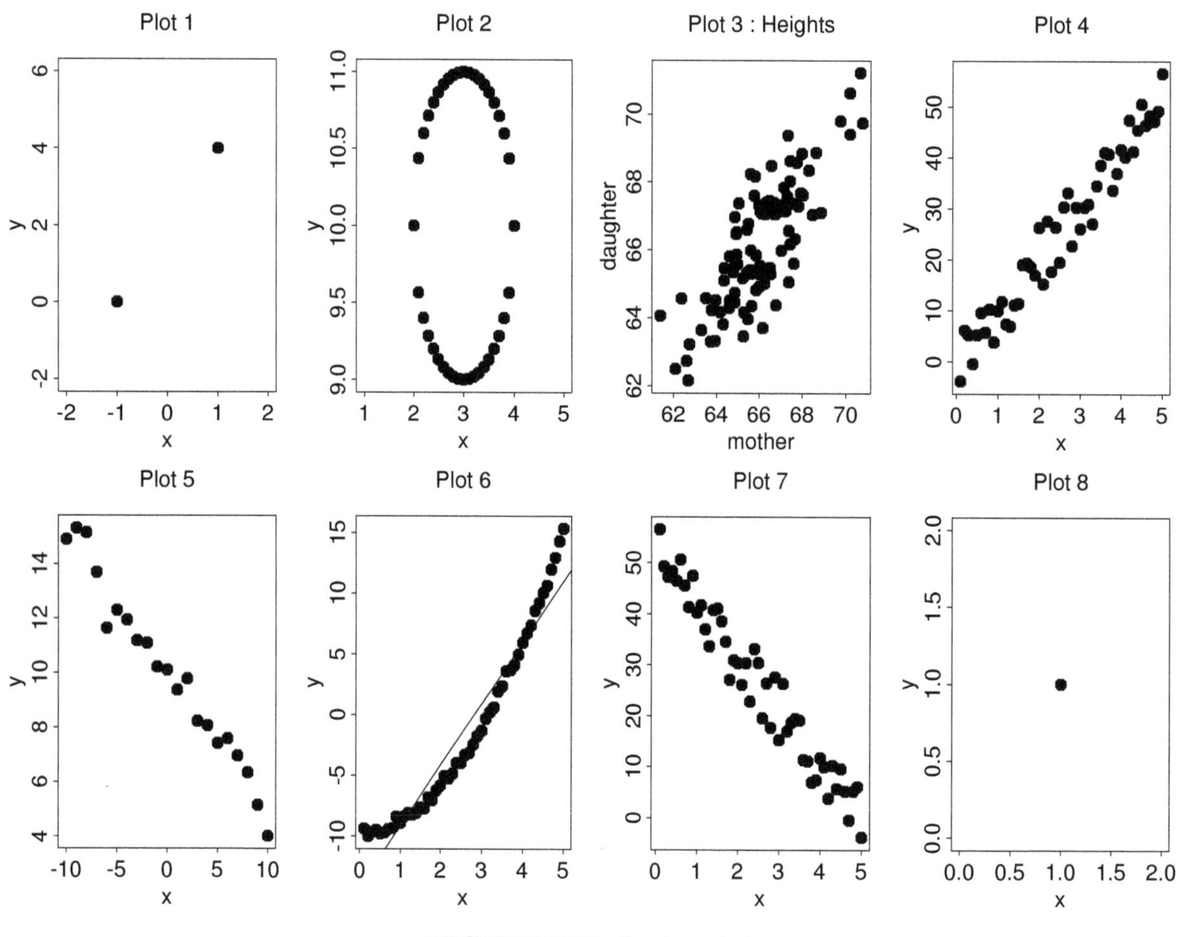

FIGURE 5.17. Scatterplots

5.1 Match the SLR statistics with one of the scatterplots shown in Figure 5.17.

a) $a = 0$, $b = 10$

b) $a = 10$, and non-zero slope

5.2 Refer to the scatterplots shown in Figure 5.17.

a) Indicate two (or more) plots where the SLR model is *not* appropriate. Explain why.

b) Two plots with identical RSquare (r^2) values.

5.3 In Plot 3 of Figure 5.17, the slope of the SLR fitted line was found to be equal to the correlation between the Mother and (adult) Daughter heights (i.e., $b = r$).

What does this imply about the (predicted) height of a daughter whose mother is *taller* than average? Explain your reasoning.

5.4 FARMPOP

The number of people living on American farms has declined steadily since 1935.
Note: **Pop** represents *millions* of persons.

a) Fit a SLR to the data

b) What are the intercept and slope estimates of the fitted SLR line?

c) Use RSquare to compute the correlation coefficient r for this dataset.

d) The intercept has a specific interpretation for this dataset.
What is the interpretation and does it make sense? Explain.

e) Use the regression equation to predict the number of people living on farms today.
Comment on your result.

5.5 ORDINAL data with entries such as *low, medium, high* do have an inherent ranking that we could code as 1, 2, 3. However, these rankings are subjective and **not** meaningful for regression analysis as the following exercise illustrates.

Here, Y = Yield of a crop and X = Coded Nitrogen levels in fertilizer.

a) Compute the correlation for Y versus X

b) If the X codes are reversed, what will the new correlation be?

5.6 ANSCOMBE Four X, Y regression datasets indexed by Plot
This exercise illustrates that numerical diagnostics alone can be quite misleading.

a) Plot each of the regression datasets.
[JMP: Put Plot in the **BY** panel of the **Fit Y by X** dialog window.]

b) Only one plot supports fitting a SLR. Explain why.

c) Fit the LS line to each plot.
   i) What is the intercept and slope for each plot? (Round to 1 decimal place)
   ii) What does the RSquare value indicate about these plots?

5.7 TRANSFORM Here we consider *transforming* the X, Y dataset

a) Plot X versus Y and fit the LS line
Use the **Linear Fit** submenu to plot the Residuals
What does the **Residual by Predicted Plot** show?

b) What transformation(s) are suggested from Figure 5.16?

c) The dataset contains the variables Y^2 and sqrtX
What **two** scatterplots appear to best linearize the dataset?
[HINT: Fit the LS line and check the residual plot for each scatterplot]

d) Argue that either transformation is adequate. [HINT: RSquare]

6

Categorical Data

Surveys and Observational Studies produce categorical data where the results are presented as percentages. Opinion polls typically include a *margin-of-error* component.

In this chapter we describe methods for analyzing categorical data.

6.1 Proportion

The simplest case involving categorical data is where an event of interest, A say, either occurs or does not. This can be generically encoded as the **indicator** variable

$$X = \mathcal{I}[A] = \begin{cases} 1 & \text{if } A \text{ occurs} \\ 0 & \text{if } A \text{ does not occur} \end{cases} \tag{6.1}$$

While this is a rather coarse classification: *"Did the patient pass the hearing test?"* it leads to research questions such as "What percentage of older adults have hearing loss?" Now the question is a little more complicated since the researcher needs to clearly specify the criteria used to define: What age is considered *older*? What are the thresholds for *loss*?

It follows that methods for analyzing a categorical variable with multiple levels will be needed. For example, the researcher may want to investigate the association between different age groups and the degree of hearing loss. We shall consider these methods later in this chapter, but first let us deal with the case of a single proportion: $p = P[X = 1]$ where p denotes the **chance** that the event A occurred. This is the *probability* $P[X = 1]$

Example 6.1 Let $X = 1$ denote the event of getting a Head from a coin toss. Assuming the coin is fair, we would intuitively think of there being a **50–50 chance** of getting a Head and set $p = P[X = 1] = 0.5$ Suppose we tossed the coin $n = 10$ times. Should we expect to get **exactly** five Heads?

Our answer to Example 6.1 is **no** in the sense that we should not be surprised if we observed getting six (6) Heads in 10 tosses. *But, what if we only observed* **one** *Head in 10 tosses?*

Question: Can we assess the fairness of the coin?

Simplify the problem to $n = 3$ tosses. The possible outcomes from (Toss 1)(Toss 2)(Toss 3) are: $HHH, HHT, HTH, THH,$ $\boxed{\text{HTT, THT, TTH}}$ $, TTT$ So the chance of getting only one H is 3 out of 8 = 0.375 (37.5%) Certainly a reasonable chance that this could occur.

However, extending this approach to $n = 10$ tosses would be **very** *tedious!*

So let's try to generalize the problem:

- For $n = 3$ tosses we had 8 possible outcomes.
- $8 = 2 \times 2 \times 2 = 2^3$ That is, each "Toss" has 2 possible outcomes H or T

  So the **total** number outcomes can be calculated by multiplying together the number outcomes for each individual Toss[1]
- If we **assume** the coin is fair (P[Head] = P[Tail] = $\frac{1}{2}$) then the chance of getting any "one" of the 8 outcome patterns is: $\frac{1}{8} = \frac{1}{2} \times \frac{1}{2} \times \frac{1}{2} = 0.125$
- There are **3 ways** to get one H in $n = 3$ tosses (boxed outcomes)
- Prob[one H in $n = 3$ tosses] $= 3 \times (0.125) = 0.375$

This suggests that for $n = 10$ tosses we have:

- $2^{10} = 2 \times 2 \times \cdots \times 2 = 1024$ Since each "Toss" has 2 possible outcomes H or T
- If we **assume** the coin is fair (P[Head] = P[Tail] = $\frac{1}{2}$) then the chance of getting any "one" of the 1024 outcome patterns is: $\frac{1}{2} \times \frac{1}{2} \times \cdots \times \frac{1}{2} = \frac{1}{1024}$
- There are **10 ways** to get **one** H in $n = 10$ tosses[2] (Toss $i = H$, all other Tosses $= T$)
- Prob[one H in $n = 10$ tosses] $= 10 \times \frac{1}{1024} = 0.0098 < 0.01$

So there is *less* than a 1% chance of getting the result of only **one** Head in $n = 10$ tosses — **"if"** the coin is **assumed** to be fair. It *can* happen, but it is **very** unlikely and we would be justified (?) in concluding that the coin is **biased** (not a fair coin). The problem is that this conclusion could be **"wrong!"** since there is, in fact, a chance (albeit small) that the result could occur with a "fair" coin.

Key Point: Statistical conclusions are NEVER absolute

---

[1] This assumes the Tosses are *"independent"* — i.e., the outcome of (Toss 1) has no influence on the outcome of (Toss 2)

[2] We can also answer questions such as "3 Heads in 10 tosses" [Answer $\binom{10}{3}2^{-10} = 120/1024 = 0.1172 \approx 12\%$]

6.2 Goodness-of-Fit (GOF)

The probabilistic approach presented for Example 6.1 works if n is small. In practice, n needs to be large to provide useful results for categorical data. (Preelection Polls typically sample $n > 1000$ respondants). The **Goodness-of-Fit** (GOF) procedure provides a useful method for assessing observed frequency distributions of categorical data versus our expectations.

Example 6.2 Let us revisit the coin toss experiment in Example 6.1, but with $n = 100$ tosses using three **different** coins. Suppose the results are as shown in Table 6.1

TABLE 6.1. Coin Toss Experiment

Toss	Coin 1	Coin 2	Coin 3	*Expected*
H	52	60	70	50
T	48	40	30	50

The *Expected* column represents what we expect (theoretically) from $n = 100$ tosses if each of the coins are assumed to be fair (P[Head] = P[Tail] = $\frac{1}{2}$) We would conclude Coin 1 is likely to be fair and Coin 3 is biased toward Heads — Coin 2 also shows a bias toward Heads, but is the bias "statistically" **significant** ? That is, what is the chance of getting a 60 – 40 split (assuming Coin 2 is fair).

For Coin 2, we begin by setting $O_i = n_i$ to be the *Observed* counts. So $O_H = 60$ and $O_T = 40$ Similarly, we define the *expected* counts E_i as $E_H = 50 = n * P[H] = 100 \times \frac{1}{2}$ and $E_T = 50 = n * P[T] = 100 \times \frac{1}{2}$ Then we compute the following quantity (with $k = 2$)

$$W = \sum_{i=1}^{k} \frac{(O_i - E_i)^2}{E_i} = \frac{(60-50)^2}{50} + \frac{(40-50)^2}{50} = 4$$

Now what? It can be shown that the statistic W has a **Pearson** Chi-squared distribution with $k - 1$ degrees of freedom (i.e., must be in one of the k groups) and so we can compute whether (or not) W exceeds the so-called **critical value** which are shown in Table 6.2.

TABLE 6.2. Chi-square critical values with right tail probability $\alpha = 0.05$

df	1	2	3	4	5	6	7	8	9	10	11	12
χ^2	3.84	5.99	7.81	9.49	11.07	12.59	14.07	15.51	16.92	18.31	19.68	21.03

Since $W = 4 > 3.84$ we would conclude Coin 2 is biased. We need JMP !

6.3 JMP NOTES for GOF

We use JMP to analyze the Coin 2 result in Table 6.1 First we need to setup the JMP dataset as shown in Table 6.3 where **Kount** is the *Observed* counts. The **Expected** column is not actually needed, but it reminds us that Coin 2 is **assumed** to be fair for the GOF analysis.

TABLE 6.3. Coin 2 data in JMP

	Coin	Kount	Expected
1	H	60	50
2	T	40	50

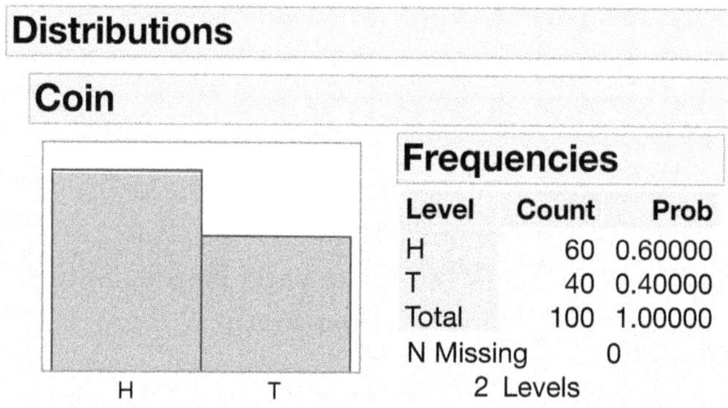

FIGURE 6.1. Coin 2 Barplot in JMP

Now let us assess the statistical "fairness" of Coin 2 using the GOF test

1. **Analyze** $\rightarrow$ **Distribution**
2. Put **Coin** as the Y, Columns and **Kount** as Freq. Click **OK**
3. From the popup by **Coin** select Test Probabilities
4. Enter the **Hypoth Prob** values 0.5, 0.5 Then click Done

When we click Done the result of the GOF analysis is shown in Figure 6.2. The **Pearson** line shows **ChiSquare** = 4.0000 which is what we calculated for W above.

6.3.1 $P-$value

Figure 6.2 also shows the quantity **Prob>Chisq** which is called a $\mathbf{p}-$**value.** Every statistical analysis method that we will encounter produces a "$p-$value" which is defined as

$P-$value = Prob[*Chance we get the Observed Result or something more* **extreme** *assuming the Expected result is true*]

Test Probabilities

Level	Estim Prob	Hypoth Prob
H	0.60000	0.5
T	0.40000	0.5

Click then Enter Hypothesized Probabilities.

Select an alternative hypothesis for testing probabilities.

- (•) probabilities not equal to hypothesized value (two-sided chi-square test)
- ( ) probability greater than hypothesized value (exact one-sided binomial test)
- ( ) probability less than hypothesized value (exact one-sided binomial test)

Done Help

Test Probabilities

Level	Estim Prob	Hypoth Prob
H	0.60000	0.5
T	0.40000	0.5

Test	ChiSquare	DF	Prob>Chisq
Likelihood Ratio	4.0271	1	0.0448*
Pearson	4.0000	1	0.0455*

Method: Fix hypothesized values, rescale omitted

FIGURE 6.2. Coin 2 GOF Dialog and Analysis in JMP

Key Point: $P-$value $< 0.05 \Rightarrow$ there is sufficient evidence **against** our assumption

That is, we "assumed" Coin 2 was fair, but the chance that is true (given the Observed result) is only 0.0455 ($< 0.05 = 5\%$) Another example is presented below.

Example 6.3 Is your stock broker better at picking winning stocks as compared to just tossing a fair coin?

Let us consider $n = 125$ brokers who picked four stocks and how many of them picked winners. The data are shown in Table 6.4 where **Pick** is the number of winning stocks picked by brokers represented in the **Kount** column and **Prop** = **Kount** / 125

TABLE 6.4. Stock Broker Picks

	Pick	Kount	Prop	p_o
1	0	3	0.024	0.0625
2	1	23	0.184	0.25
3	2	51	0.408	0.375
4	3	39	0.312	0.25
5	4	9	0.072	0.0625

Could we have done just as well as the brokers by simply tossing a fair coin four times? For example, the chance that none ($X = 0$) of our picks were winning stocks would be the same as tossing four "Tails" or $P[X = 0] = P[T] * P[T] * P[T] * P[T] = (0.5)^4 = 1/16 = 0.0625$ Similarly $P[X = 4] = 1/16$ (all winners), $P[X = 1] = 4/16 = P[X = 3]$ (one winner out of four — which has the same chance as one *loser* out of four). Lastly, $P[X = 2] = 6/16$ since there is six ways to pick two winners: $WWLL, WLWL, WLLW, LWWL, LWLW, LLWW$ (We could also use the Binomial distribution). Hence, the **p-o** is our "Null" hypothesis that we want to compare to the observed performance of the brokers in **Prop**

It can be seen that the brokers did do *better* since they picked a higher proportion of two or more winners. From a practical perspective this may suggest using a broker can be beneficial (ignoring trade fees). The statistical question is whether they did *significantly* better than tossing a coin. This can be done using the GOF test.

GOF Test Let k be the total number of categories to which n_i subjects are uniquely observed in category $i = 1, 2, \ldots, k$ and $n = \sum n_i$ be the total number of subjects. Then the GOF test of $H_o : p = p_{o_i}$, $i = 1, 2, \ldots, k$ where the p_{o_i} are user specified is given below.

Set $O_i = n_i$ and define the *expected* counts $E_i = n * p_{o_i}$ Then under H_o the GOF test statistic is

$$W = \sum_{i=1}^{k} \frac{(O_i - E_i)^2}{E_i} \sim \chi^2_{k-1}$$

provided all the $E_i > 0$ and no more than 20% of the $E_i \leq 5$

In the stock picking example, $k = 5$ and $O_i =$ **Kount** We will use JMP to compute the GOF test as follows. The JMP output is shown in Figure 6.3.

1. **Analyze → Distribution**
2. Put **Pick** as the Y, Columns and **Kount** as Freq. Click **OK**
3. From the popup by **Pick** select Test Probabilities
4. Enter the **p-o** values Then click **Done**
5. The p−value for the "Pearson" test is the GOF test result.

From the Pearson row in Figure 6.3, the GOF test result is $W = 7.6080$ with $k = 5$ Thus, $\text{df} = k - 1 = 4$ and the p−value = **Prob>Chisq** = 0.1070 > 0.05 Statistically, this suggests brokers are no better at picking winning stocks than if one simply tossed a fair coin.

6.4 Contingency Tables

When subjects are cross-classified by two factors, A and B say, a *Contingency Table* analysis can be performed to test the hypothesis H_o : *no association between A and B*. Synonymous terms are *Crosstabs* and *Two-way* analysis

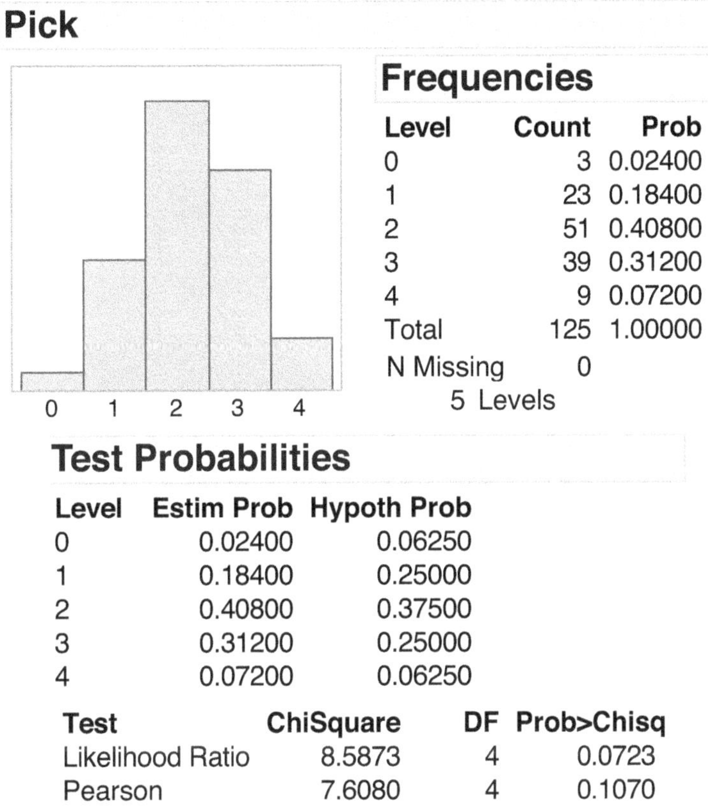

Frequencies

Level	Count	Prob
0	3	0.02400
1	23	0.18400
2	51	0.40800
3	39	0.31200
4	9	0.07200
Total	125	1.00000
N Missing	0	
5 Levels		

Test Probabilities

Level	Estim Prob	Hypoth Prob
0	0.02400	0.06250
1	0.18400	0.25000
2	0.40800	0.37500
3	0.31200	0.25000
4	0.07200	0.06250

Test	ChiSquare	DF	Prob>Chisq
Likelihood Ratio	8.5873	4	0.0723
Pearson	7.6080	4	0.1070

Method: Fix hypothesized values, rescale omitted

FIGURE 6.3. Stock Broker Picks Analysis

Example 6.4 Is there Gender bias in Hiring? Consider the case where 100 candidates were interviewed for a position at a large company. Of the 60 Male applicants 15 were Hired, and of the 40 Female applicants, 10 were Hired.

Clearly, we could ask many questions about this "scenario" such as why there were more Male applicants (i.e., did the company use a different screening profile for Females? This potentially might suggest gender bias). For simplicity, let us pursue the statistical analysis and address the contextual concerns in our conclusion. From the information we can represent the study as a two-way table shown in Table 6.5.

In order to compare rows and columns we will need to convert to proportions since the marginal totals are different. Thus we see that $15/60 = 0.25 = 10/40$ so Males and Females were Hired in the *same* proportion. It would seem that gender bias was **not** a precondition in the hiring process of this company (notwithstanding our contextual concerns). But what if the data was represented as shown in Table 6.6.

Now we see that 50% of Female applicants were Hired versus only $\sim$ 8.3% of Males and so gender bias **does** appear to be a precondition in the hiring process of this company. That

TABLE 6.5. Hiring versus Gender. Example 1

		Hiring Status		
		Not Hired	Hired	
Gender	Male	45	15	60
	Female	30	10	40
		75	25	100

TABLE 6.6. Hiring versus Gender. Example 2

		Hiring Status		
		Not Hired	Hired	
Gender	Male	55	5	60
	Female	20	20	40
		75	25	100

is, the applicant's "Gender" is associated with whether they were hired or not. So how do we assess this "association" statistically? To do this we need to consider H_o : *no association between A and B* which is equivalent to saying A and B are "independent." If this statement is true then any cell count should only depend on the marginal row and column counts (and total sample size). For example, if you are a Male applicant your "chance" of being hired should be no different from a Female applicant. Based on Table 6.5 we see that

$$P[Male\ and\ Hired] = \frac{15}{100} = P[Male] \times P[Hired] = \frac{60}{100} \times \frac{25}{100}$$

and the *expected* number of Males hired is $(n = 100) \times P[Male\ and\ Hired] = 15$ Thus we can employ the GOF approach where we compare the Observed cell count to the Expected cell cell (based on the marginal/total counts). Table 6.7 shows the general $r \times c$ case of a contingency table for Factors A (r levels) and B (c levels)

TABLE 6.7. Contingency Table $r \times c$ case

		Factor B				
		B_1	B_2	$\cdots$	B_c	
	A_1	n_{11}	n_{12}	$\cdots$	n_{1c}	$n_{1.}$
Factor A	A_2	n_{21}	n_{22}	$\cdots$	n_{2c}	$n_{2.}$
	$\vdots$	$\vdots$	$\vdots$	$\ddots$	$\vdots$	$\vdots$
	A_r	n_{r1}	n_{r2}	$\cdots$	n_{rc}	$n_{r.}$
		$n_{.1}$	$n_{.2}$	$\cdots$	$n_{.c}$	$n_{..}$

Thus the *Expected* cell count is $E_{ij} = \frac{\text{row i total} \times \text{column j total}}{\text{sample size}}$ Thus

$$W = \sum_{\text{cells}} \frac{(O_{ij} - E_{ij})^2}{E_{ij}} \sim \chi^2_{(r-1)(c-1)}$$

is our test statistic. Below we will perform the JMP analysis of three different Commercials versus their Ratings as shown in Table 6.8 from a total of 900 respondents. The format needs to be changed so that the Ratings are stacked in a single column for each Commercial.

TABLE 6.8. Advertisement Ratings

	Commercial	very favorable	favorable	neutral	unfavorable	very unfavorable	Total
1	A	32	87	91	46	44	300
2	B	53	141	76	20	10	300
3	C	41	93	67	36	63	300
4	Total	126	321	234	102	117	900

The result is shown in Table 6.9 using the following steps in JMP.

1. Delete the row and column **Total**
2. **Tables** → **Stack** and select the Rating columns. Enter into `Stack Columns`
3. Rename `Stacked Data Column` as **Kount** and `Stacked Label Column` as **Rating** Click **OK**
4. In the new stacked dataset click on the **Ratings** header.
5. **Cols** → **Standardize Attributes** then `Column Properties` → `Value Ordering` and shift `very favorable` to the top of the list. Click **OK**

The * beside **Rating** shows that the attributes of this variable have been modified from the JMP default which would list the categories in alphabetical order in the analysis report. Now we can use **Analyze** → **Fit Y by X** to obtain the contingency table analysis shown in Figures 6.4 and 6.5.

The mosaic plots are a visual tessellation of the cell frequencies with the width of the horizontal Rating categories representing the column Total responses in Figure 6.4. The vertical heights within each Rating category represent the row percentage which is the fourth cell entry in each row of Figure 6.4. We note that a significant association between Commercial and Rating would be suggested by different tile heights across commercials which is evident in the unfavorable Rating categories. Specifically, Commercial B (green) received the fewest unfavorable and very unfavorable Ratings compared to A (red) and C (blue). Alternatively, the mosaic plot can be plotted with Commercials on the horizontal axis.

The Pearson statistic (our W) shows there is a statistically significant ($p-$value $< .0001$) association between Commercial and the respondents Ratings. Thus we can proceed to interpret the association. Note that if the Pearson $p-$value was *not* significant, we should not

TABLE 6.9. Commercial Ratings Reformatted

ex4-ads-done.jmp

ex4-ads-done...
Source

Columns (3/0)
Commercial
Rating *
Kount

Rows
All rows 15
Selected 0
Excluded 0
Hidden 0
Labelled 0

	Commercial	Rating	Kount
1	A	very favorable	32
2	A	favorable	87
3	A	neutral	91
4	A	unfavorable	46
5	A	very unfavorable	44
6	B	very favorable	53
7	B	favorable	141
8	B	neutral	76
9	B	unfavorable	20
10	B	very unfavorable	10
11	C	very favorable	41
12	C	favorable	93
13	C	neutral	67
14	C	unfavorable	36
15	C	very unfavorable	63

proceed with a post-hoc analysis from a statistical perspective. However, there may be trends that are worth noting even when the result is not statistically significant.

We will use the conditional proportions to compare between Ratings or across Commercials These are the 3rd and 4th entries in each cell from the default JMP contingency table output. If we condition on commercial, then we are comparing the `Col %` cell entries (3rd) down each column. Thus, Commercial B received 17.67% very favorable and 47.00% favorable Ratings which are clearly larger than the corresponding Ratings for Commercials A and C. If we compare Commercials within each Rating category, again we see that Commercial B dominates the very favorable (42.06%) and favorable (43.93%) Ratings.

So which comparisons should be used (row or column)? The answer is both provide useful information, but it is likely that the company producing these commercials will be more interested in comparisons between A, B, and C (columns) except that the 53.85% very unfavorable Rating versus 32.54% very favorable for Commercial C in the row comparison is worth noting (respondents either loved or hated Commercial C). So herein lies the problem with 2-way contingency tables: other factors are not accounted for. For example, did younger respondents prefer Commercial C versus older?

Key Point: Contingency tables only show *association* not causation.

Table 6.6 suggests Gender "discrimination" but that would imply Gender alone accounted for the hiring practice of the company which is causation (with legal ramifications).

Contingency Analysis of Commercial By Rating

Mosaic Plot

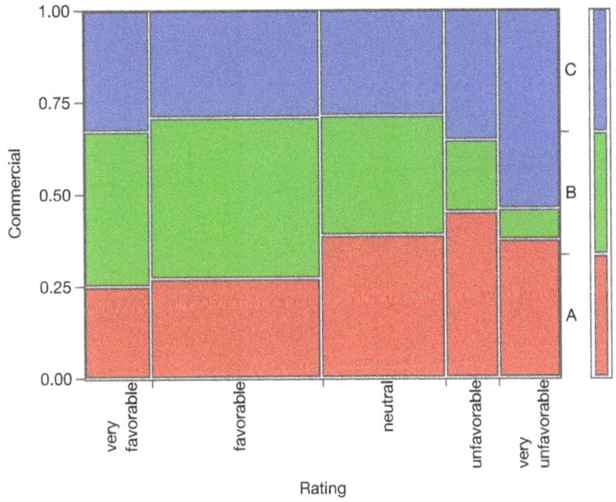

1.00
0.75
0.50
0.25
0.00
Commercial
C
B
A
very favorable
favorable
neutral
unfavorable
very unfavorable
Rating

Freq: Kount

Contingency Analysis of Rating By Commercial

Mosaic Plot

1.00
0.75
0.50
0.25
0.00
Rating
very unfavorable
unfavorable
neutral
favorable
very favorable
A
B
C
Commercial

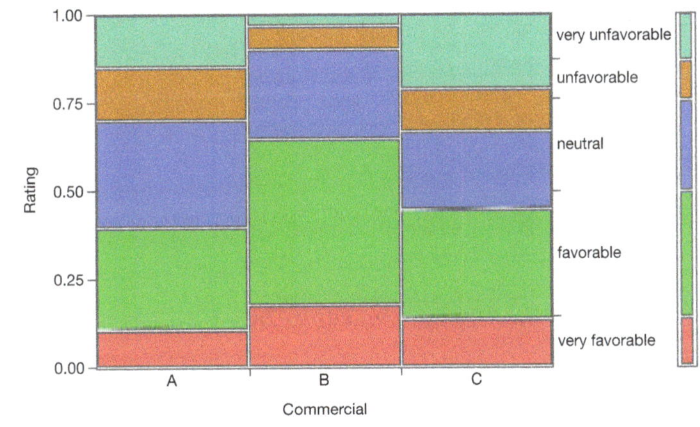

Freq: Kount

FIGURE 6.4. Commercial Ratings: Mosaic Plots

Contingency Analysis of Commercial By Rating

Contingency Table

Rating	Count / Total % / Col % / Row %	A	B	C	Total
		Commercial			
	very favorable	32	53	41	126
		3.56	5.89	4.56	14.00
		10.67	17.67	13.67	
		25.40	42.06	32.54	
	favorable	87	141	93	321
		9.67	15.67	10.33	35.67
		29.00	47.00	31.00	
		27.10	43.93	28.97	
	neutral	91	76	67	234
		10.11	8.44	7.44	26.00
		30.33	25.33	22.33	
		38.89	32.48	28.63	
	unfavorable	46	20	36	102
		5.11	2.22	4.00	11.33
		15.33	6.67	12.00	
		45.10	19.61	35.29	
	very unfavorable	44	10	63	117
		4.89	1.11	7.00	13.00
		14.67	3.33	21.00	
		37.61	8.55	53.85	
	Total	300	300	300	900
		33.33	33.33	33.33	

Tests

N	DF	-LogLike	RSquare (U)
900	8	39.631468	0.0401

Test	ChiSquare	Prob>ChiSq
Likelihood Ratio	79.263	<.0001*
Pearson	72.521	<.0001*

FIGURE 6.5. Commercial Ratings: Contingency Table

6.5 Simpson's Paradox

The contrast between the *Hiring versus Gender* examples presented in Tables 6.5 and 6.6 illustrate that a 2-way contingency table analysis may **not** necessarily provide useful results. That is, for Table 6.5 we agreed that Gender did not appear have an impact on Hiring, whereas for Table 6.6 there seemed to be a clear Gender bias which showed Females were much more likely to be Hired than Male candidates.

The problem is that we "know" Hiring status (should) depend on the qualifications of the candidate and requirements for the Job position and not on the candidate's gender. If the Job position required an MBA degree, but candidates were **not** prescreened then the Gender "association" evident in Table 6.6 may be completely irrelevant.

This suggests including another factor, C say, may (or may not) change the *direction* of association between the original CT analysis of A versus B. We consider the following example using Death Penalty data from Radelet (1981).

Example 6.5 The Death Penalty convictions (Yes or No) were recorded based on the race of the Defendant[3] and the race of the Victim. The results are presented in Table 6.10. Does the *direction* of association change when the Victim's race is excluded?

TABLE 6.10. Death Penalty convictions versus Race

Defendant Race	Victim Race	Death Penalty Yes	Death Penalty No	Total	Percentage Yes
White	White	19	132	151	12.6
	Black	0	9	9	0.0
Black	White	11	52	63	17.5
	Black	6	97	103	5.8
Ignoring Victim Race					
White	—	19	141	160	**11.9**
Black	—	17	149	166	**10.2**

Thus **controlling** for the Victim's race (e.g., White) we see that the percentage who received a "yes" death penalty verdict was higher for Black defendants (17.5) than for White defendants (12.6). This *direction* of association also holds when the Victim's race is Black (5.8 versus 0.0).

However, when the Victim's race is ignored in the combined CT, the direction of association is **reversed**: White defendants who received the death penalty verdict (11.9) was higher than for Black defendants (10.2). This type of reversal is called **Simpson's paradox**.

Simpson's paradox An association that holds between A, B, and C reverses direction when the data is combined to A and B only.

Key Point: Relevant covariates (if available) should be included in a CT analysis.

As we have seen, the direction of the *partial* associations across all the levels of the controlling factor become "hidden" when absorbed into the combined CT and can result in a reversal of the direction of the *marginal* association.

[3]The 326 subjects were defendants in homicide indictments in 20 Florida counties during 1976–1977.

Chapter 6 Exercises

6.1 BIRTHS versus day of the week dataset. Here, we are interested in whether there is significant difference between the number of births by day of the week based on the GOF test of no difference. Three separate samples are given in the dataset.

a) What probability is assumed if there is no difference between birth and day?

b) Hence, what the *Expected* number of births per day for the Birth1 sample?

c) Perform the GOF test for the Birth1 sample. What is your conclusion?

JMP: A shortcut to Test Probabilities is to switch the radio button to Fix omitted ... and enter "1" for all the Hypoth Prob boxes Then click Done

d) Show the other two Birth datasets are significant. Explain why day would be a factor.

6.2 Responses from 500 CEO's who were asked how satisfied they were in their present position and how they rated their own performance are shown in the table below.

	Job Satisfaction	
Performance	Want to Leave	Not Paid Enough
Really Good	75	150
Even Better	50	225

a) If a CEO was randomly selected from this group, find the probablity that the person:

i) Rates their performance as really good.

ii) Feels they are not paid enough *given* they rated their performance as even better.

b) If $n = 5$ additional CEO's were surveyed and their responses added to the table above, how many *different* 2×2 tables are possible? [HINT: Balls in Urns problem]

c) Pick any cell and show that independence does *not* hold.

6.3 Survival rates at two different Hospitals is shown in the Table below.

a) Compute the death rate at each Hospital when the patient is in **Bad** condition

b) Compute the death rate at each Hospital when the patient is in **Good** condition

c) Show the results above are reversed if patient condition is ignored.

d) This is an example of Simpson's Paradox. Explain the contradiction in these results.

Hospital Attended	Patient's Condition	Patient Outcome		
		Died	Survived	Total
A	Good	6	594	600
	Bad	57	1443	1500
B	Good	8	592	600
	Bad	8	192	200

6.4 MARRIAGE In a National Longitudinal Survey, one question Male and Female teens were asked was *"What are the chances you think you will be married in the next 10 years?"*

a) Create the CT for this dataset.
   Put Marriage as Y, Gender as X and Kount as Freq
   Explain what the significant Pearson result suggests
b) What percentage of Females responded Definitely?
c) Discuss the conditional distribution of Gender for each Marriage category.
   What is your conclusion about Male and Female opinions on future marriage?

6.5 EDUCATION Years on Fist Job versus Years of Education

a) Perform the CT analysis
b) What do the results indicate ? Provide details.

6.6 SHOPPING Shopping preference (see Note) versus Income level
Note: Other represents anything other than Online or Supermarket
[e.g., Convenience store, Local farmer markets, Discount stores, etc.]

a) Perform the CT analysis (Put Y = Shop)
b) The Pearson test is highly significant [**Prob>ChiSq**]. What does this suggest?
c) Eliminate the Other category and redo the CT analysis.
d) What do the results in part (c) indicate? Provide details.

6.7 Suppose you have a 3×4 CT
Explain why chi-square test is not valid in the following situations.

a) The **Expected** count for cell (1,3) is 0 (zero)
b) The marital status (single, married, widowed, divorced) was recorded for three (3) age groups (young, middle-aged, older). There were $n = 10$ subjects in each age group.
c) A survey question had four (4) options where participants from three (3) age groups could select *all that apply.*

6.8 ACTIVITY Physical activity versus Fruit consumption was studied for a sample of 1184 college students. The premise for the study was that physical activity and fruit consumption may decline as students transition from high school to college.

a) Show the CT analysis is significant.
b) Discuss what the conditional distributions suggest.

Part III

Statistical Inference

7
Sampling Distributions

In section 4.8.1 we introduced the Central Limit Theorem (CLT) which we restate here.

Theorem: Central Limit Theorem (CLT)
Let $X_1, X_2, \ldots, X_n$ be a *random sample* from a process X having mean μ and variance σ^2 and $\overline{X}$ be the mean of the random sample. Then, as n increases, the following random variable converges in distribution to a standard Normal random variable.

$$Z = \frac{X - \mu}{\sigma/\sqrt{n}} \to N(0, 1)$$

Remarks:

1. Knowledge of the underlying pdf of the process X is not required.
2. In general, the approximation becomes very good for $n > 30$
   [The result is exact for any n if the process $X \sim N(\mu, \sigma^2)$]
3. By *unstandardizing* it follows that $\overline{X} \to N(\mu, \frac{\sigma^2}{n})$

The importance of the first remark is that knowledge of the *actual* distribution of the population is **not** needed. The distribution can be discrete or continuous provided the sample size is sufficiently large ($n > 30$).

Problem: The third remark would appear to make the CLT result "irrelevant" since it requires we know the true population mean μ and standard deviation σ That is, how can we use the CLT if we do not know μ and σ ?

7.1 Confidence Interval (CI)

Let us consider the problem of finding z^* such that $P[-z^* < Z < z^*] = 0.95$ where $Z \sim N(0,1)$ That is, we want to find z^* so that exactly 95% of the area under the standard Normal curve (Figure 4.8) lies in an interval centered around zero.

This implies 5% lies *outside* the interval from $-z^*$ to z^* By symmetry of the $N(0,1)$, it follows that $2.5\% = 0.0250$ of the area lies *below* $-z^*$ or, equivalently, $0.9750 = 0.95 + 0.0250$ lies below z^* From Table A2, Appendix B, we find that $z^* = 1.96$

Hence $0.95 = P[-1.96 < Z < 1.96]$

Now we can work backward by letting $Z = (\overline{X} - \mu)/(\sigma/\sqrt{n})$ be the standardized random variable obtained from $\overline{X}$. By the CLT, $Z \sim N(0,1)$ assuming $n > 30$ Since standardization preserves area we have:

$$\begin{aligned} 0.95 &= P[-1.96 < Z < 1.96] \\ &\approx P[-1.96 < \frac{\overline{X} - \mu}{\sigma/\sqrt{n}} < 1.96] \\ &= P[-1.96\frac{\sigma}{\sqrt{n}} < \overline{X} - \mu < 1.96\frac{\sigma}{\sqrt{n}}] \\ &= P[\overline{X} - 1.96\frac{\sigma}{\sqrt{n}} < \mu < \overline{X} + 1.96\frac{\sigma}{\sqrt{n}}] \end{aligned} \tag{7.1}$$

This result is exact for any n if the sample came from a Normal population. Otherwise, the CLT "approximation" is very good provided n is sufficiently large $(n > 30)$.

7.1.1 CI for μ with σ Known

We can interpret the last equation (7.1) as an (approximate) 95% *Confidence Interval* (CI) for an unknown population mean μ Hence, if we **pretend** that we "know" σ, this can be computed from an observed random sample as the 95% CI estimate for μ:

$$\overline{x} \pm 1.96\frac{\sigma}{\sqrt{n}} \tag{7.2}$$

Specifically, a CI provides an *interval estimate* for μ and we can objectively assign a level of *confidence* such as 95% which is the (theoretical) probability that the true population mean lies in the interval. The distinction from a *point estimate* for μ is important since we have no way of knowing how close $\overline{x}$ is to μ without the measure of variability SE $=\sigma/\sqrt{n}$

Caution: It is important to note that the "95% CI" we compute from our observed random sample only represents a "single" realization of the event $\overline{X} \pm 1.96\frac{\sigma}{\sqrt{n}}$ defined in (7.2). Hence, it is "technically" incorrect to say *"we are 95% confident μ lies in this* **(particular)** *CI"* because the term "95% confident" refers to the *expected* probability from an ensemble of *all* CI realizations. For our purposes, we will interpret the *"95% CI for μ"* pragmatically, with the implicit understanding that the computed CI is an *estimate.*

In practice, the default confidence level used is 95% which corresponds to $z^* = 1.96$. This can be changed by using a different confidence level such as 99% or 90% and the effect on the CI is summarized below:

- 99% corresponds to $z^* = 2.576$ so the CI is **wider** (less precise), but there is less *risk* (1%) that μ is **not** in the CI
- 90% corresponds to $z^* = 1.645$ so the CI is **narrower** (more precise), but there is greater *risk* (10%) that μ is **not** in the CI
- The only other way to increase the precision of a CI is to increase the sample size n

7.1.2 Margin of Error

This is the quantity $\boxed{m = z^* \frac{\sigma}{\sqrt{n}}}$

Thus, given any three of the components m, z^*, σ, n the other component can be obtained. In particular, we are often interested in determining the sample size n needed for a specified margin of error.

Example 7.1 Find n for a 95% CI with $m = 0.14$ and $\sigma = 4$ Solving for n gives:

$$n = \left(\frac{z^*\sigma}{m}\right)^2 = \left(1.96(4)/0.14\right)^2 = 3136$$

Choosing a small m can make the required sample size n *very* large! If an estimate of σ is not available (e.g., this is a new study), then we can specify $m = \delta\sigma$ where δ is some fraction such as 0.5 or 0.2 etc. [The σ cancels out in the formula for n]

7.2 Hypothesis Test (HT)

Inference for μ based on a CI can be thought of as "data-orientated" in the sense that the sample data are simply used to compute the interval $\overline{x} \pm z^*\text{SE}$ Then we use the CLT to *infer* this interval will contain μ with probability equal to the confidence level determined by z^* However, many studies have a specific objective or *directional hypothesis* that the investigator is interested in "testing." Consider the following scenarios.

1. Drug A *will* lower LDL in patients with high cholesterol
2. Taking a Kaplan course *will* improve your SAT test score.
3. The water bottle *does* contain 500ml

In the first two cases, the question of interest is "by how much?" and in the third case, is the amount specified correct. Of course, the actual change will vary between individuals in cases 1 and 2, and not "every" water bottle will contain exactly 500ml. That is, these claims are based on the **average** $\overline{x}$ from a sample and it is then *inferred* the claim appears to valid (or not) for the population mean μ of *all* subjects in the target population.

Key Point: Our *inference* applies to the population mean μ — **not** individuals

In statistical terms for scenario 1, we are *hypothesizing* that the LDL level μ for all high cholesterol patients who take Drug A will be lower — on **average** — than the average, μ_o say, for all high cholesterol patients who do not take Drug A. Equivalently, we could state this in the form $H_A : \mu < \mu_o$ meaning that, on average, *all* patients will see a reduction in LDL regardless of their high cholesterol level after taking Drug A. Similarly, for the Kaplan example, $H_A : \mu > \mu_o$ (score increases, on average, for all candidates taking the Kaplan course), and $H_o : \mu = 500$ (all water bottles contain 500ml, on average).

Key Point: The value of μ_o needs to **specified** for a HT

For scenario 3 we *can* intuitively "test" this by comparing the sample average of the amount of water contained in n bottles to 500ml. So if $\overline{x}$ is close to 500ml then the claim would seem to be valid. Similarly, scenarios 1 and 2 could be tested using the "null" hypothesis $H_o : \mu = \mu_o$ (Drug A and the Kaplan course make no difference, on average).

7.2.1 HT for μ with σ Known

We use the **4 Step Process** introduced in Chapter 2 to analyze the Drug A cholesterol scenario. To do this, we need some more information. Suppose a random sample of $n = 36$ patients with LDL levels over 150 were studied and prescribed Drug A. After 3 months the LDL levels of these patients were measured again and resulted in $\overline{x} = 141$. Assume $\sigma = 18$

STATE Does Drug A reduce LDL (on average) in high cholesterol patients?

PLAN Here we take $\mu_o = 150$

Null Hypothesis: $H_o : \mu = \mu_o$ represents "no change" Drug A was not effective

Alternative Hypothesis: $H_A : \mu < \mu_o$ Drug A reduced average LDL levels

SOLVE We need to perform the following:

1. Check **conditions**: a) Random sample is claimed; b) $n > 30$ so HT should valid (if no outliers); c) $\sigma = 18$ is known
2. Compute the **test statistic**: $z_o = \dfrac{(\overline{x} - \mu_o)}{\sigma/\sqrt{n}} = \dfrac{(141 - 150)}{18/\sqrt{36}} = -3$
3. Find the p–value: $p\text{–value} = P[Z < z_o] = 0.0013 < 0.05$ so Reject H_o

CONCLUDE There is (statistical) evidence that Drug A reduced LDL levels *on average*

Remark: Statistical evidence is **not** necessarily the same as practical (medical) significance. Note that a healthy LDL level for adults is less than 100 mg/dl so a 9 point average reduction might suggest Drug A was not very effective from a medical perspective.

7.2.2 $P-$value for μ with σ Known

Definition: $P-$value $= P[$getting z_o or something **more extreme** assuming H_o is true$]$

Key Point: What is regarded as "more extreme" depends on H_A

For the Kaplan scenario $H_A : \mu > \mu_o \Rightarrow p-\text{value} = P[Z > z_o]$

For the water bottle scenario $H_A : \mu \neq 500$ which implies that we were looking to see if there was change in *either direction* from $\mu_o = 500$ Hence something more extreme would be anything **below** $-|z_o|$ or **above** $|z_o|$ (even though there is only a single z_o computed).

Thus, for a two-sided alternative $H_A : \mu \neq \mu_o \Rightarrow p-\text{value} = 2 * P[Z < -|z_o|]$

Remarks:

1. The CI and HT based on $z-$procedures assume σ is **known**. This assumption is somewhat unrealistic, and we will address this in chapter 8.
2. We can extend $z-$procedures to 2-sample problems e.g., $H_o : \mu_x = \mu_y$. Again, we would need to assume σ_x and σ_y are known
3. JMP can perform the 1-sample $z-$test

7.3 Population Proportion

Recall the Bernoulli random variable defined in section 4.7.1 where X only takes the value 0 or 1 and $p = P[X = 1]$ is the "success" probability. Let $q = (1 - p) = P[X = 0]$ Then

$$\mu = p \quad \text{and} \quad \sigma = \sqrt{pq}$$

Now let $X_1, X_2, \ldots, X_n$ be a **random sample** from the "population" X Hence

$$\hat{p} = \overline{X} = \frac{\sum X_i}{n} = \frac{\text{sum of the number of successes}}{\text{total sample size}}$$

It follows that we can use the CLT to derive a CI and HT for the true proportion p

7.3.1 CI for p

Substituting $\hat{p} = \overline{x}$ and $\sqrt{pq} = \sigma$ in $\overline{x} \pm \sigma/\sqrt{n}$ we have:

A CI for p is given by

$$\hat{p} \pm z^* \sqrt{\frac{pq}{n}} \tag{7.3}$$

Key Point: The difference here is that the SE $= \sqrt{pq/n}$ in (7.3) **depends** on p

In (7.2) σ was an *anciliary* quantity (independent of μ) and the CLT only depended on σ known and $n > 30$ Here, we also require the following condition in order to use the CLT

$$np \geq 5 \quad \text{and} \quad nq \geq 5 \tag{7.4}$$

where we can use $\hat{p}$ to estimate whether condition (7.4) holds.

Similarly, we cannot use the true (unknown) p in the SE, so we need to estimate it. We actually have two options:

1. Use $\hat{p}$ so that $SE_{est} = \sqrt{\hat{p}\hat{q}/n}$
2. Use $p = 0.5$ which gives the **maximum** value $SE_{max} = \sqrt{0.5(1-0.5)/n} = 1/(2\sqrt{n})$

7.3.2 Margin of Error for Surveys

The margin-of-error using p^* ($\hat{p}$ or 0.5) is given by

$$m = z^* \sqrt{\frac{p^* q^*}{n}} \quad \Rightarrow \quad n = (z^*/m)^2 p^* q^*$$

Example 7.2 Polls often report a 3% margin-of-error. Assume a 95% CI. Compute n

$$\text{Using } p^* = 0.5 \qquad n = (1.96/0.03)^2(0.5(1-0.5)) = 1037.1 \nearrow 1038$$

Key Point: Proportion analysis (such as Polls) require **large** samples to be useful

7.3.3 HT for p

This turns out to be easier than the CI analysis since $H_o : p = p_o$ determines p Hence

$SE_o = \sqrt{p_o q_o/n} \Rightarrow z_o = (\hat{p} - p_o)/SE_o$

$P-$values are determined as before by the H_A

Remarks:

1. The CI and HT for p based on $z-$procedures generally require larger sample sizes since condition (7.4) needs to be satisfied. That is, the success and failure events cannot be too rare.
2. We can extend $z-$procedures to 2-sample proportions e.g., $H_o : p_x = p_y$. However, these can be better analyzed using Contingency Tables
3. JMP can perform a 1-sample CI for p A GOF test can be used for a HT

7.4 Distributions Related to the Normal

There are several distributions that are used extensively in statistical analysis and can be derived directly in terms of the Normal distribution. For convenience we define them now and leave discussing their properties until the appropriate context for their use arises.

Chi-square: χ^2_ν

Let $Z \sim N(0,1)$ then the random variable $W = X^2$ has a Chi-square distribution with $\nu = 1$ degree of freedom, denoted by $W \sim \chi^2_1$.

The sum of r independent standard Normals (squared) results in a Chi-square distribution with $\nu = r$ degrees of freedom. A Chi-square random variable W has already played an important role in Chapter 6 for analyzing the significance of categorical data. It can also be used to analyze variance-type quantities.

Student's t: t_ν

Let $Z \sim N(0,1)$ and $W \sim \chi^2_\nu$ be independent. Then the random variable

$$T = \frac{Z}{\sqrt{W/\nu}} \sim t_\nu$$

has a Student's t distribution with ν degrees of freedom. Standardizing a Normal random variable requires knowledge of σ^2. The t distribution provides the correct pdf when the sample variance s^2 is used to estimate σ^2. This is used extensively in Chapter 8

F Disribution: $\mathcal{F}_{\nu_1,\nu_2}$

Let $W_1 \sim \chi^2_{\nu_1}$ and $W_2 \sim \chi^2_{\nu_2}$ be independent Chi-square random variables. Then the following random variable has an F distribution with ν_1, ν_2 degrees of freedom.

$$F = \frac{W_1/\nu_1}{W_2/\nu_2} \sim \mathcal{F}_{\nu_1,\nu_2}$$

To compare the relative difference between two variance-type quantities, the ratio of two independent Chi-square random variables can be used. The resulting pdf of this ratio has an F distribution. This is used for ANOVA methods discussed in Chapter 10.

Chapter 7 Exercises

7.1 A new car model claims $\mu = 25$ mpg (miles per gallon) with $\sigma = 2$ mpg Assume mpg is Normally distributed for this car model.

a) Let $\overline{X}$ be the sample mpg average from $n = 16$ cars. What is the distribution of $\overline{X}$?

b) Hence compute the chance that $\overline{X}$ is within ± 1 mpg of the claim.

c) If you bought this car, would you expect to get the same mpg range as above? Explain.

7.2 Assume the average weekly earnings from a small business follow a Normal distribution with $\mu = \$61,000$ and $SE = \$3,000$. Use this information to answer the following questions.

a) i) 95% of average weekly earnings will be in what interval centered at μ?
   ii) An exceptionally good week would imply that average weekly earnings exceeded what value?

b) What is the chance that the average weekly earnings is between \$52,900 and \$66,040?

c) Explain why the shorter interval in part (a)(i) is *better* than the part (b) interval
   [NOTE: The length of an interval is: (Upper limit) – (Lower limit)]

7.3 RADON A study of 12 radon detectors that were exposed to a controlled rate of 105 picocuries per liter of radon over three days. Assume Normality and $\sigma = 9$

a) Compute the 2-sided p–value for $H_o : \mu = 105$
   JMP: Radon → Test Mean Enter 105 for the mean and 9 for the true stdev

b) Based on your HT above, is there significant evidence at the 5% level that the mean reading differs from the true value of 105? Explain your reasoning.

c) How many detectors would the researchers have needed to be 95% sure their sample mean was within ± 1 picocuries per liter of the true mean radon reading?

7.4 Let $\mu = 8.8, \sigma = 1.0$ be the heartbeats **per five seconds** for adult runners. Let $\overline{X}$ be the heartbeat rate based on as random sample of 25 runners

a) i) What is the (approximate) distribution of $\overline{X}$ using the CLT?
   ii) Do you think this approximation is reasonable? Explain.

b) Use part (a)(i) to compute $P[\overline{X} < 8.5]$

c) What is the probability that the average heartbeat of a runner will be less than 100 beats per minute? [HINT: Convert $X = 100$ per minute to $\overline{X}$ per 5 seconds]

7.5 Let $\hat{p} = 0.05$ be the proportion of defectives from a random sample of $n = 100$ items.

a) Show that the CLT condition holds.

b) Use SE_{est} to compute a 95% CI for the true defective rate.

c) Show the margin-of-error is $m^* = 0.098$ if we use SE_{max}

d) i) If the manufacturer wanted $m^* = 0.01$ what sample size is required?
   ii) Explain why this result is not surprising.
   [HINT: Magnitude of the reduction in m^*]

8 Inference

The assumption that σ is "known" is unrealistic and the obvious solution would be to simply replace σ by the sample standard deviation s. Hence, a 95% CI for μ becomes:

$$\overline{x} \pm 1.96 \frac{s}{\sqrt{n}}$$

This is referred to as a *large sample* approximation which can be used for large $n > 50$. The main problem is that now there is additional variability due to s which depends on the sample selected. Hence, even if we are sampling from a Normal population, this CI is too narrow in general for smaller sample sizes. In this case, the correct sampling distribution for $\overline{X}$ follows a $t-$distribution.

8.1 One-Sample $t-$Procedures

Definition: $t-$**distribution** Let $X_1, X_2, \ldots, X_n$ be a *random sample* from a $N(\mu, \sigma^2)$ process with sample mean $\overline{X}$ and sample standard deviation S Then, the random variable

$$T = \frac{\overline{X} - \mu}{S/\sqrt{n}} \sim t_{n-1}$$

has a $t-$distribution with $n-1$ *degrees of freedom (df)*

Remarks:

1. The assumption of Normality is required for the result to be exact.
2. The pdf (density curve) of a $t-$distribution depends on the *df* parameter.

3. $T \to N(0,1)$ as the *df* parameter increases.

Example 8.1 From Table B, Appendix B, we have the following

95% CI with $n = 8 \Rightarrow df = 7$ and $t^* = 2.3646$

95% CI with $n = 21 \Rightarrow df = 20$ and $t^* = 2.0860$

99% CI with $n = 5 \Rightarrow df = 4$ and $t^* = 4.6041$

95% CI with $n = 67 \Rightarrow df = 66$ which is not listed in Table B.
Thus, we use the *smaller* $df = 60$ entry where $t^* = 2.0003$

Example 8.1 shows that the $t-$distribution is *heavy tailed* compared to the N(0,1) in that t^* is always further away from zero than the corresponding z^* quantile ($df = \infty$) This is illustrated in Figure 8.1 where we have also included the $t_{2.3}$ curve. That is, where the $df = 2.3$ parameter involves a fractional part. While this does not make sense for the random variable T defined above (can't have a sample of size $n = 3.3$), the t_k density curve can be defined mathematically as a continuous function of k so the $t_{2.3}$ is simply a density curve that lies between the t_2 and t_3 curves. We will encounter fractional degrees of freedom with the *two sample* $t-$procedures discussed later in this chapter.

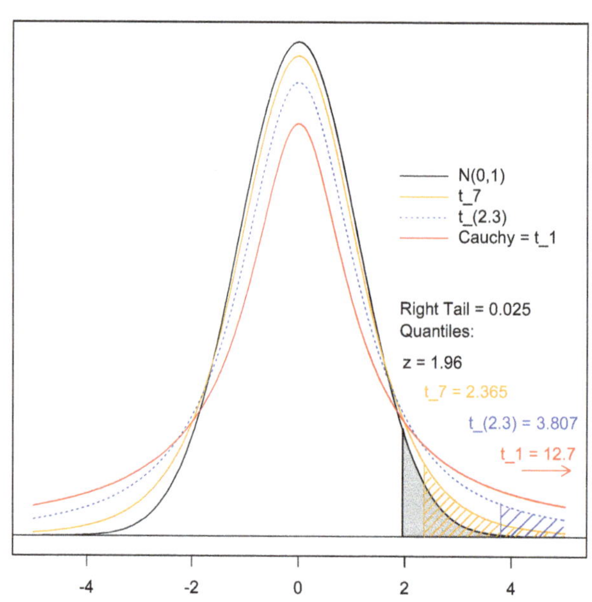

FIGURE 8.1. t_k density curves vs. $N(0,1)$

t−Interval for μ

Given a random sample of size n from a Normal population, a CI with confidence level C is given by

$$\overline{x} \pm t^* \frac{s}{\sqrt{n}}$$

where t^* is the t_{n-1} quantile for confidence level C

8.1.1 t−Tests

The general principles of Hypothesis Testing can be illustrated in the case of the mean of a process. We will use the 4-Step method illustrated in Chapter 2. To add context to the discussion, consider the following example.

Example 8.2 Suppose a computer company, MIRFT say, wants to assess its problem-to-resolution program. MIRFT has a diverse range of products so the database of cases is quite large ($n > 40$) and updated with monthly profiles.

STATE This assessment concerns the impact of a retraining course that was recently completed by the customer service staff. MIRFT would like to know if the retraining resulted in a *significant* reduction in resolution time. (Was the retraining cost-effective?)

PLAN Statistically speaking, the result that MIRFT wants to test is:

$$H_o : \mu = \mu_o \qquad \text{versus} \qquad H_a : \mu < \mu_o$$

where μ_o is the expected duration of problem-to-resolution cases prior to retraining and μ is the average duration for customer resolution by the staff after retraining. Note that the "test" is symbolically represented in terms of *two* hypotheses which are referred to as:

H_o **Null** Hypothesis. $H_o : \mu = \mu_o$ is usually an expression of the status quo (retraining has no effect), and our decision to reject should only happen when there is *significant* evidence to the contrary.

H_A **Alternative** Hypothesis. Here we have a *left-tailed-test* (LTT) $H_a : \mu < \mu_o$ since MIRFT is only interested in whether the retraining "reduced" customer resolution times

NOTE For the Kaplan example above, we would use a *right-tailed-test* (RTT) $H_A : \mu > \mu_o$ If "no" direction is implied by the context of the study, we use the **default** *two-tailed-test* (2TT) $H_A : \mu \neq \mu_o$

SOLVE: Because we employ a random sample to perform our HT, the decision we make about the process may be incorrect. Since we can never be certain that H_o is True or False, our *inference* or conclusion about the process is always phrased in terms of "sufficient evidence ..." (to reject or accept H_o). Thus we need to quantify what we mean by "sufficient" For the moment, assume that we know the truth but carry out the HT anyway. Our decision can then be classified into one of the following four situations:

Decision	H_o is True	H_o is False
Accept H_o	✓	Type II Error
Reject H_o	Type I Error	✓

If H_o is True, we can control the "risk" of making an incorrect decision (Type I Error). This is called the *significance level* α which, by convention, is set as 0.05. That is, we agree that if the chance of H_o being True is 5% or less based on our sample evidence, then we will consider this to be *significant* evidence against H_o. The $p-$value discussed below is a more informative measure of Type I Error.

By convention, the significance levels that are accepted in practice are:

5% $\alpha = 0.05$ This is the default level. We interpret our test as significant at the 5% level if the chance of rejecting H_o assuming it is True, is less than 5%

1% $\alpha = 0.01$ Rejecting H_o at this level is referred to as *strongly* significant.

10% $\alpha = 0.10$ Rejection at this level is interpreted as *weak* evidence against H_o.

If H_o is False, then quantifying our decision would be difficult since we do not know the actual value of the process mean μ. What we do expect however, is that our "test" procedure should become more *powerful* as μ gets further away from the hypothesized value μ_o. That is, the chance of making an incorrect decision (Accept H_o when it is False) decreases (rapidly) as $|\mu - \mu_o|$ increases.

Test Statistic

For a HT involving μ, our standard test statistic is based on the $t-$distribution for $\overline{X}$. The observed value of our test statistic is

$$t_o = \frac{\overline{x} - \mu_o}{s/\sqrt{n}}$$

Under H_o the sampling distribution associated with t_o is t_{n-1} and software such as JMP will provide the $p-$value associated with t_o.

$P-$value

The $p-$value of a HT is the *probability of getting your test statistic — or something more extreme — assuming H_o is true.* The $p-$value approach for HT has two advantages

- Comparing the $p-$value to 0.05 works for *any* test. $p-$value $< 0.05 \Rightarrow$ significance
- The $p-$value shows how likely H_o is true.
  Smaller $p-$value $\Rightarrow$ more evidence against H_o

There are clearly differences between $p-$values such as 0.001, 0.047, 0.051, 0.07, and 0.34. Table 8.1 gives $p-$value qualifications that can be used to express "significance"

TABLE 8.1. $P-$value Qualifications

Decision	$P-$value	Qualification	Statement
Accept H_o	≥ 0.05	> 0.10	No evidence
		$0.06 - 0.10$	Weak evidence
		$0.05 - 0.06$	Marginal non-significance
Reject H_o	< 0.05	$0.04 - 0.05$	Marginal significance
		$0.01 - 0.04$	significant evidence
		< 0.01	Strong evidence

Decision

Having computed the observed value of the test statistic and the $p-$value, our "test" consists of checking if the $p-$value < 0.05 If so, then we "Reject H_o," otherwise we "Accept H_o"

CONCLUDE

Though we may be finished with the statistical analysis, it is still necessary to express our "decision" as a conclusion *in the context of the problem.*

Our conclusion will be of the form "there is *[in]*sufficent evidence at the $100\alpha\%$ level to suggest that retraining does *[not]* reduce the duration of problem-to-resolution cases"

The *[ ]* parts correspond to the **accept** H_o decision.

Robustness

The requirement that the random sample comes from a Normally distributed population would seem to limit the usefulness of a $t-$interval for inference about an unknown population mean. Intuitively, however, we should expect that the assumption of Normality will become less important as the sample size increases since $T \to N(0, 1)$ and the fact that the CLT provides a very good approximation for $n > 30$ in general (irrespective of the distribution of the population). This is indeed the case, and the one-sample $t-$procedure can be shown to be *distributionally robust* to the assumption of Normality as n increases.

Guidelines: For the one-sample $t-$procedure to be considered reliable, the following guidelines can be used. Reliability here means that the actual confidence level of the CI computed using the $t-$interval will be close to the C selected. Equivalently, the actual significance level of a HT using the $t-$test will be close to the α selected (usually 0.05).

$n < 15$	Normal distribution. Histogram has "center" peak and tails off to both sides. No outliers. Realistically, the histogram will generally not be useful for small samples and the assumption of normality may need to be argued from *context* of the data. Otherwise, a nonparametric method (section 8.4) can be used and compared to the $t-$procedure result.
$15 \leq n < 40$	Skewness becomes increasingly acceptable as n gets larger. That is, the histogram should be approximately "bell-shaped" if n is close to 15, but can be more skewed as n approaches 40. No outliers.
$n \geq 40$	Strong skewness and other non-normal distributions are acceptable. No outliers in the sense that such point(s) are inconsistent with the underlying distribution of the data.

8.2 Two-Sample $t-$Procedures

There are two basic design structures for the two-sample situation which are illustrated in the schematic below. For the independent samples case, two distinct populations are involved which are assumed to be independent and random samples of size n_x and n_y are selected. In the paired samples case, a "single" population is measured twice under two different conditions and the *differences* between each subject's response is used for analysis.

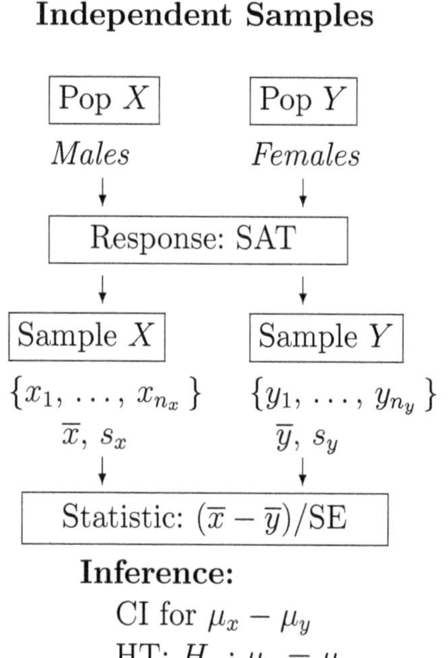

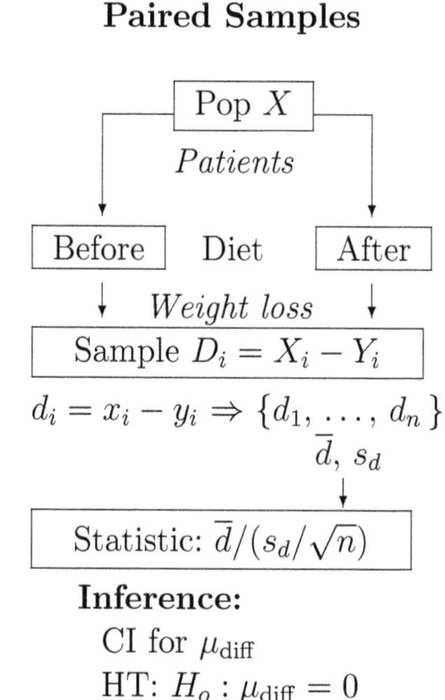

8.2.1 Paired $t-$procedure

The "Before and After" type of study illustrated in the design schematic above is used to control for an individual's *baseline* which can vary greatly between individuals within each condition. However, we would also expect their before state to be correlated to their after state. Using the individual *differences* between the Before and After states accommodates both the baseline effect and expected correlation. Thus, the analysis reduces to the One-sample $t-$procedures discussed in the previous section. Specifically:

CI for μ_{diff} $\quad \bar{d} \pm t^* \frac{s_d}{\sqrt{n}}$ $\quad$ where t^* is the t_{n-1} quantile for a confidence level C

HT for $H_o : \mu_{\text{diff}} = 0$ $\quad t_o = \frac{\bar{d}}{(s_d/\sqrt{n})} \sim t_{n-1}$ $\quad$ under H_o

Key Point: The interpretation depends on how the *differences* are calculated

For example, in the Weight Loss example indicated in the Paired Samples schematic, we would interpret a significant *positive* result as providing sufficient evidence that the Diet worked: the subjects lost weight on average. However, if we had calculated $d_i = y_i - x_i$ then a significant *negative* result would provide sufficient evidence of the same conclusion.

Double-Blind Experiment

The "Before and After" design is often used for measuring human responses in Placebo experiments: A patient is first given Drug A and then later Drug B (or vice versa), one of which is a Placebo. To ensure the validity of the results — controlling for "interview" or patient bias — neither the patient *or* the nurse/doctor administering the drug knows which drug is which. (Somebody has to know of course!) Such a procedure is called a "Double-Blind" experiment.

Matched Pairs

A variation on the Paired Samples design is where the observations are collected from two distinct groups, but the observations have a natural connection or "match." For example, voting preferences of a husband and wife, performance on the SAT by a brother and sister from the same family. Again, we expect correlation between these "pairs" and so differences are also used to used to analyze a "matched pairs" design.

Equivalence Tests

Equivalence tests assess whether two different (but related) products, drugs, or methods can be considered (statistically) "equivalent" or not. We use the same hypothesis testing approach, but change the null and alternate hypothesis so that the null represent that the two systems are different (change) and now the alternate is that they are the same within a specified tolerance interval. [What we call a CI]

Example 8.3 The before and after results from eight patients taking a pain relief tablet versus placebo was conducted. Four randomly selected patients started with the placebo, and then after a wash-out period, were switched to the active ingredient drug. The reverse administration was used for the other four patients. The data and analysis is shown in Figure 8.2. Pain relief was measured in hours.

	Patient	Placebo	Drug	Diff (P-D)
1	1	10	8	2
2	2	3	1	2
3	3	11	9	2
4	4	4	2	2
5	5	11	6	5
6	6	6	6	0
7	7	7	6	1
8	8	12	10	2

Summary Statistics

Mean	2
Std Dev	1.4142136
Std Err Mean	0.5
Upper 95% Mean	3.1823121
Lower 95% Mean	0.8176879
N	8

Test Mean

Hypothesized Value	0
Actual Estimate	2
DF	7
Std Dev	1.41421

	t Test
Test Statistic	4.0000
Prob > |t|	0.0052*
Prob > t	0.0026*
Prob < t	0.9974

FIGURE 8.2. Paired $t-$Test

The context of the study implies that we expect the Drug to provide pain relief faster, on average, compared to no treatment (Placebo). That is, $\mu_{\rm P} > \mu_{\rm D}$ which corresponds to the average difference $\mu_{\rm diff} > 0$ where $\text{diff}_i = \text{P}_i - \text{D}_i$ The appropriate test is therefore $H_o : \mu_{\rm diff} = 0$ versus $H_A : \mu_{\rm diff} > 0$ and from the JMP output, the appropriate $P-$value is `Prob` $>$ `t` $= 0.0026$ which is significant. We can conclude from this study that there is strong evidence that the Drug provides faster pain relief, on average.

Note that the 95% CI for $\mu_{\rm diff}$ is $(0.82, 3.18)$ which is positive and does *not* contain zero. Hence, at the 5% significance level ($\alpha = 0.05$), we can also conclude that the Drug provides significantly faster pain relief. Of course, statistical significance is not necessarily the same as practical significance so we cannot say whether or not the 2 hour average reduction in pain relief is medically meaningful.

8.2.2 Independent Samples

The paired design requires more careful monitoring in that the patient needs to be in similar condition for both the before and after phases for the example above. In fact, the study described in Example 8.3 is a crossover design with a washout period that is needed to avoid a "carry over" effect. However, one advantage of a paired design is that fewer subjects are needed compared to an independent two-sample design.

The two-sample study illustrated in the schematic is obviously needed when Males are randomly selected since it would be somewhat impractical to observe the same individual's SAT performance as a Female! The point, of course, is that two-sample designs are required in many situations and are relatively simple to implement. However, the $t-$procedure used for the analysis of a two-sample sample is a little more complicated than one might expect.

Let $\mu_x, \overline{x}$ and s_x be the population mean and sample statistics for Population X. Then, assuming the population X is normally distributed (or the sample size n_x is sufficiently large as defined by the one-sample guidelines), we have that:

$$T_x = \frac{\overline{x} - \mu_x}{(s_x/\sqrt{n_x})} \sim t_{n_x - 1} \qquad \text{and similarly} \qquad T_y = \frac{\overline{y} - \mu_y}{(s_y/\sqrt{n_y})} \sim t_{n_y - 1}$$

with the same conditions applying to Population Y. This suggests we could investigate μ_x versus μ_y based on the (univariate) parameter $\mu_x - \mu_y$ and the combined statistic T will have a sampling distribution t_k with degrees of freedom $k = n_x - 1 + n_y - 1 = n_x + n_y - 2$

$$T = \frac{(\overline{x} - \overline{y}) - (\mu_x - \mu_y)}{\text{SE}} \sim t_k \qquad \text{where} \quad \text{SE} = \sqrt{\frac{s_x^2}{n_x} + \frac{s_y^2}{n_y}}$$

Unfortunately, this is **not** the case unless it is assumed $\sigma_x^2 = \sigma_y^2$ The assumption of equal population variances is tenuous at best. If the two groups are homogeneous, this assumption is probably reasonable, but it is clearly not applicable to every two-sample study.

Key Point: T can be approximated by a t_k distribution with k computed as below

It can be shown the following value of k provides an optimal t_k approximation for T provided both sample sizes n_x and n_y are greater than 5.

$$k = \frac{\left(\frac{s_x^2}{n_x} + \frac{s_y^2}{n_y}\right)^2}{\frac{1}{n_x - 1}\left(\frac{s_x^2}{n_x}\right)^2 + \frac{1}{n_y - 1}\left(\frac{s_y^2}{n_y}\right)^2}$$

This approximation is computed by JMP and often k will have a fractional component. Without software, the conservative value of $k = \min(n_x, n_y) - 1$ can be used. Thus

CI for $\mu_x - \mu_y$ $\quad (\overline{x} - \overline{y}) \pm t^*\text{SE}$ $\quad$ where t^* is the t_k quantile for confidence level C

HT for $H_o : \mu_x = \mu_y$ $\quad t_o = \dfrac{\overline{x} - \overline{y}}{\text{SE}} \sim t_k$ under H_o

Example 8.4 Using the same data as the paired study, but assuming the eight patients in each group are independent, the two-sample t–test results are shown in Figure 8.3.

The two-sample t–test result is *not* significant `Prob > t` = 0.1240 and the 95% CI for $\mu_{\text{P}} - \mu_{\text{D}}$ is $(-1.56, 5.56)$ which contains zero. This is because the large variation within each group effectively "masks" what we found to be a significant difference in the paired design.

CI versus HT Inference

In two-sample studies, interest is usually on whether there is a significant difference between the mean response of the two treatment groups. That is, $H_o : \mu_x = \mu_y$ is a conceptually "natural" (null) hypothesis as compared to a CI for $\mu_x - \mu_y$. In the one-sample case, the

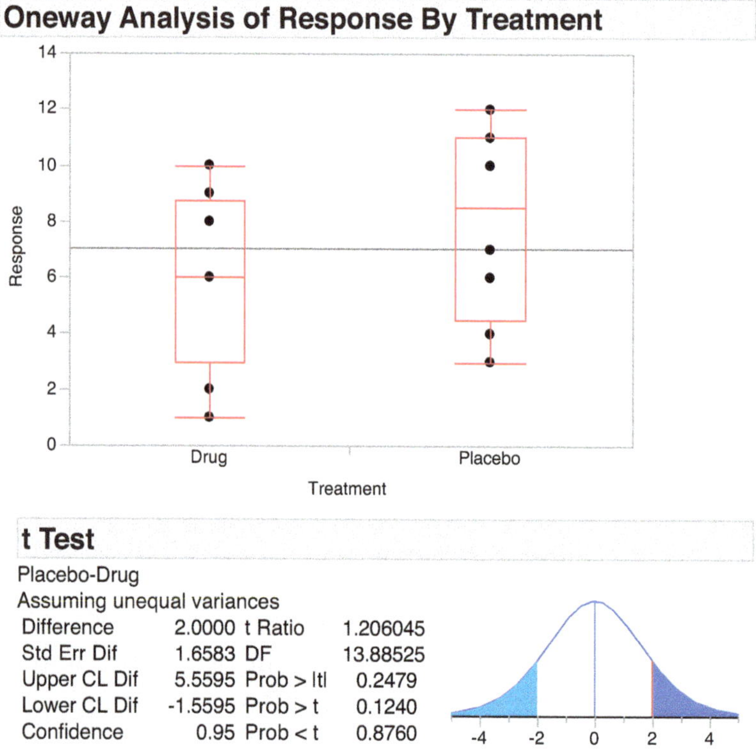

Difference	2.0000	t Ratio	1.206045
Std Err Dif	1.6583	DF	13.88525
Upper CL Dif	5.5595	Prob > |t|	0.2479
Lower CL Dif	-1.5595	Prob > t	0.1240
Confidence	0.95	Prob < t	0.8760

FIGURE 8.3. Independent $t-$Test

CI inference approach would be used when the analysis is data-orientated: "where does the data suggest μ lies?" versus "I think $\mu = \mu_o$."

Pooled $t-$Procedure

If we are prepared to assume $\sigma_x^2 = \sigma_y^2$ (equal population variances) and that both samples come from independent Normal distributions, then

$$T = \frac{(\overline{x} - \overline{y}) - (\mu_x - \mu_y)}{\text{SE}_{\text{pooled}}} \sim t_{n_x + n_y - 2}$$

where

$$\text{SE}_{\text{pooled}} = \sqrt{\frac{(n_x - 1)s_x^2 + (n_y - 1)s_y^2}{n_x + n_y - 2}\left(\frac{1}{n_x} + \frac{1}{n_y}\right)}$$

While the CLT deals with the Normality assumption, we do not recommend using the pooled $t-$procedure in practice since there is no way for us to "truly" verify the equal population variances assumption. It can be shown that the pooled $t-$procedure becomes increasingly less reliable as ratio σ_x^2/σ_y^2 gets further away from 1. Software is readily available to provide the optimal t_k approximation.

Robustness

The two-sample $t-$procedure is actually more "robust" in the sense that results can be considered valid whenever both sample sizes n_x and n_y are greater than 5. However, some guidelines are worth noting:

1. $n_x = n_y$ Equal sample sizes mean that the same amount of information is used to compute the sample statistics from each group. If we "know" (or suspect) one group may have higher variability, then oversampling can be used.

2. $\hat{f}(x) \sim \hat{f}(y)$ Ideally X and Y should have the same distribution SHAPE. That is, both symmetric, or both skewed in the same direction. Again, if the sample sizes are greater than 5 then we can regard the two-sample $t-$procedure as valid irrespective of the population distributions.

3. The two-sample $t-$procedure is **not** robust against outliers and requires larger sample sizes when large variations within each group is to be expected.

8.3 Variance Procedures

Inference for a population variance σ^2 can be based on the sample variance

$$S^2 = \frac{1}{n-1}\sum_{i=1}^{n}(X_i - \overline{X})^2$$

which can be shown to be an **unbiased** estimator of σ^2 for fixed n. That is, $E[S^2] = \sigma^2$ When the underlying population distribution is $N(\mu, \sigma^2)$ the sampling distribution of S^2 follows a Chi-squared distribution with $n-1$ degrees of freedom. Specifically,

$$W = \frac{(n-1)S^2}{\sigma^2} \sim \chi^2_{n-1}$$

Like the t-distribution, the Chi-square quantiles w^* are only tabulated for certain tail areas α and degrees of freedom k where $P[W > w^* = \chi^2_k(\alpha)] = \alpha$ The important difference is the Chi-square distribution is positive $(W > 0)$ and right skewed so quantiles for both tails are given. Figure 8.4 shows some Chi-square density curves.

Example 8.5 From Table C, Appendix B, we have the following

$\alpha = 0.975$ and $n = 10 \Rightarrow df = 9$ and $w^* = \chi^2_9(0.975) = 2.0879$

$\alpha = 0.025$ and $n = 10 \Rightarrow df = 9$ and $w^* = \chi^2_9(0.025) = 19.0228$

$\alpha = 0.99$ and $n = 2 \Rightarrow df = 1$ and $w^* = \chi^2_1(0.99) = 1.571(-4) = 0.0001571$

$\alpha = 0.025$ and $n = 37 \Rightarrow df = 36$ which is not listed in Table C. Thus, we use the *smaller* $df = 30$ entry $w^* = \chi^2_{30}(0.025) = 46.979$

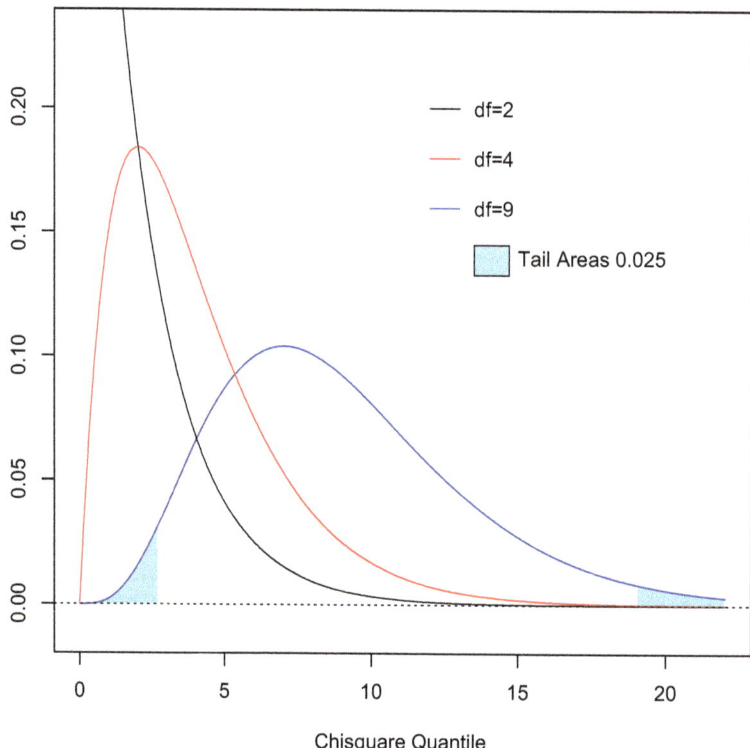

FIGURE 8.4. Chi-square Density Curves: χ^2_{df}

Example 8.6 A random sample of size $n = 10$ from a Normal distribution with unknown variance σ^2 resulted in $s^2 = 5.4$ Then $W = (n-1)S^2/\sigma^2 \sim \chi^2_9$ and the w^* quantiles that define the left and right tail areas = 0.025 shown in Figure 8.4 are:

$$w_l = \chi^2_9(0.975) = 2.700 \quad \text{and} \quad w_u = \chi^2_9(0.025) = 19.023 \quad \text{so that}$$

$$0.95 = P[w_l < W < w_u] = P\left[w_l < \frac{(n-1)S^2}{\sigma^2} < \mathbf{w_u}\right] = P\left[\frac{(n-1)S^2}{\mathbf{w_u}} < \sigma^2 < \frac{(n-1)S^2}{w_l}\right]$$

Hence, a 95% CI for σ^2 is $[9(5.4)/19.023\,,\, 9(5.4)/2.700] = [2.555, 18.000]$

Remarks

1. By taking square roots, the 95% CI for σ in the example above is: $[1.598, 4.825]$
   Note: This is default CI computed by JMP when **Confidence Interval** is selected.
2. Since W is right-skewed the upper bound of the CI for σ^2 can be quite large
3. Since $\sigma^2 > 0$, a more precise upper bound can be obtained using a **one-sided** CI $[0, (n-1)s^2/w_l^*]$ where $w_l^* = \chi^2_{n-1}(0.95)$ (right tail area = 0.05)
4. Under $H_o : \sigma^2 = \sigma_o^2$ it follows that $W_o = (n-1)S^2/\sigma_o^2 \sim \chi^2_{n-1}$ provides the appropriate test statistic when the population is Normal.

Example 8.7 JMP output for σ^2_{diff} of the pain relief data in Figure 8.2 is shown below.

Diff (P-D)

Confidence Intervals

Parameter	Estimate	Lower CI	Upper CI	1-Alpha
Mean	2	0.817688	3.182312	0.950
Std Dev	1.414214	0.935041	2.878309	0.950

Confidence Intervals

Parameter	Estimate	Lower CI	Upper CI	1-Alpha
Mean	2	1.052711	2.947289	0.900
Std Dev	1.414214	0.997611	2.541555	0.900

Test Standard Deviation

Hypothesized Value	1
Actual Estimate	1.41421
DF	7

Test	ChiSquare
Test Statistic	14
Min PValue	0.1024
Prob < ChiSq	0.9488
Prob > ChiSq	0.0512

FIGURE 8.5. Chi-square CI and HT for σ^2_{diff}

JMP expresses the CI for σ^2 in terms of **Std Dev** so the 95% CI for σ_{diff} is $[0.935, 2.878]$ The lower **Confidence Intervals** report shows that the upper bound = 2.542 for a **one-sided** 95% CI for σ can be obtained from the 90% CI report.

In the **Test Standard Deviation** report we are actually testing $H_o : \sigma^2 = (\sigma^2_o = 1)$ so the Test Statistic $= W_o = (n-1)s^2/\sigma^2_o = 14$ The $P-$value for a right-sided alternative $H_A : \sigma^2 > 1$ is `Prob > ChiSq` = 0.0512 which suggests there is marginally nonsignificant evidence that $\sigma^2_{\text{diff}} = 1$

Comparing Two Variances

Let S^2_x and S^2_y be the sample variances from random samples of size n_x and n_y obtained from independent Normal populations X and Y say. Then

$$W_x = \frac{(n_x-1)S^2_x}{\sigma^2_x} \sim \chi^2_{n_x-1} \qquad \text{and} \qquad W_y = \frac{(n_y-1)S^2_y}{\sigma^2_y} \sim \chi^2_{n_y-1}$$

From section 7.4 it follows that the ratio

$$F = \frac{W_x/(n_x-1)}{W_y/(n_y-1)} = \frac{S^2_x/\sigma^2_x}{S^2_y/\sigma^2_y} \sim \mathcal{F}_{n_x-1,n_y-1}$$

where $\mathcal{F}_{n_x-1,n_y-1}$ is the $F-$distribution with **numerator** degrees of freedom $n_x - 1$ and **denominator** degrees of freedom $n_y - 1$ Like the chi-square distribution, the F density curves are right skewed and are only defined for positive quantiles.

Our main interest will be related to use of F for hypothesis tests in applications where $H_o : \sigma^2_x = \sigma^2_y$ versus $H_A : \sigma^2_x > \sigma^2_y$. In this case, the appropriate "test" statistic is:

$$F_o = S^2_x/S^2_y$$

and our inference concerns whether the right-sided $P-$value = `Prob > F` provides sufficient evidence to conclude that the variation in population X is significantly greatly than the variation of population Y This will be dealt with in more detail in later chapters.

Remark: So far, we have focused mainly on inferential procedures for LOCATION estimates with the appropriate SE to account for the inherent variability associated with a random sample. However, there are situations where the direct study and analysis of SPREAD is important, such as in the *quality improvement* of a product. For example, if you buy a new car, computer or software, appliance, etc., your expectation is that it "works" as claimed by the manufacturer. Thus, *improvement* is achieved by reducing the variability of the components that make up a product.

8.4 Nonparametric Procedures

The inference procedures we have discussed to this point are called **parametric** in that they assume the random sample comes from a Normal distribution. The CLT (Central Limit Theorem - chapter 7 and section 4.8.1) provides distributional robustness as the sample size n increases. However, outliers, small samples, and skewness can adversely affect the $t-$procedure and result in misleading conclusions.

Let us begin with the location model

$$X_i = \theta + \epsilon_i \quad , \quad i = 1, 2, \ldots, n$$

where $\{X_i\}$ is the random sample and θ is the population parameter of interest. For the one-sample $t-$procedure $\theta = \mu$ and ϵ_i represents the (unobserved) ERROR term which is assumed to have a Normal distribution.

Question: How can we avoid the Normality assumption?

We can use a **nonparametric** procedure which converts the $\{X_i\}$ data into "Ranks" which is distribution-free and can be regarded as valid regardless of the underlying distribution of the population from which we select our random sample. We discuss "Ranks" in detail in section 8.4.2. We will start with the Sign Test where $\theta = median$

Caution: Nonparametric procedures are less **efficient**[1] than the $t-$procedure if the random sample $\{X_i\}$ came from a process that has a Normal distribution

8.4.1 Sign Test

When the distribution is observed (or known) to be strongly skewed, heavy tailed, discrete, or has outliers and the sample size is not large enough for us to use the one sample $t-$procedure with reasonable confidence, the **sign test** provides a simple alternative. It can be shown the

[1]Efficiency is a measure that allows us to compare two different test procedures in terms of sample size and SE. This is beyond the scope of this book.

Sign Test is actually "more" efficient than the $t-$test procedure (see Ott and Longnecker 2001 Table 5.7, page 248) when the distribution is heavy tailed and/or strongly skewed.

Example 8.8 Let $H_o : \theta = 10$ versus $H_A : \theta > 10$ be the HT of interest where θ is the population *median.* For the random sample below, we assign a positive sign "+" if the observed sample value is greater than the hypothesized median $\theta_o = 10$, otherwise we assign the sign is "−"

Data	11	10	20.3	80.4	25	16	47	75	11	987	$\# > \theta_o$
Sign	+	−	+	+	+	+	+	+	+	+	9

Is there sufficient evidence to support that the median has increased from $\theta_o = 10$?

Given that 9 out 10 of the observed values were greater than $\theta_o = 10$ it would seem intuitively reasonable to conclude that the population *median* has increased. Can we quantify our intuition ? Note that if H_o is true, then $P[X \leq \theta_o] = P[X \geq \theta_o] = 0.5$ by definition of the *median.* Hence, we can express $M_o = \#\{X_i > \theta_o\} = \sum_{i=1}^{n} \mathcal{I}[X_i > \theta_o]$ where the indicator function $\mathcal{I}[x > \theta_o] = 1$ or 0 according to whether $x > \theta_o$ or $x \leq \theta_o$, respectively.

Key Point: $M_o \sim Bin(n, p = 0.5)$ under $H_o : \theta = \theta_o$

This does **not** depend on the distribution of X_i and thus the test based on M_o is called *nonparametric* or *distribution free.* Recall that the $p-$value for a test is defined as:

$P-$value $= P[$ obtaining the test statistic M_o or something *more extreme,* given H_o is true $]$

For the example above, $M_o = 9$ and something more extreme would be $M_o = 10$ (more evidence *against* the null hypothesis in favor of $H_A : \theta > 10$). Thus,

$$p\text{−value} = P[M_o = 9, 10] = \left[\binom{10}{9} + \binom{10}{10}\right]\left(\tfrac{1}{2}\right)^{10} = \tfrac{11}{1024} = 0.011 < 0.05$$

which provides significant evidence that the population median is greater than 10. In contrast, the right-sided $t-$test for $H_o : \mu = 10$ is *not* significant ($p-$value = `Prob>t` = 0.1241) due to the extreme outlier.

8.4.2 Wilcoxon Signed-Rank Test

A more efficient procedure that is provided in JMP and other software is the Wilcoxon Signed-Rank Test. This assumes the underlying distribution is *symmetric* and uses the (signed) **Ranks** of the data. For the Sign Test example (after sorting) we have:

Data	10	11	11	16	20.3	25	47	75	80.4	987
Rank*	1	2	3	4	5	6	7	8	9	10
Rank	1	2.5	2.5	4	5	6	7	8	9	10

where Rank $= i$ is assigned to i'th largest data value. In the event of "ties" (same data values), it is common practice to average the initial Rank* assigned to the ties.

The signed rank test for the mean model $H_o : \theta = \mu_o$ uses the **magnitude** of the deviations $|X_i - \mu_o|$ to determine the overall Ranks, then computes the test statistic

$$W_o = \sum_{i=1}^{n} \text{sgn}(X_i - \mu_o)\text{Rank}\{\,|X_i - \mu_o|\,\}$$

Example 8.9 Let us consider $H_o : \theta = 30$ for the data above.

Data	10	11	11	16	20.3	25	47	75	80.4	987
$\lvert X_i - 30\rvert$	20	19	19	14	9.7	5	17	45	50.4	957
Rank	7	5.5	5.5	3	2	1	4	8	9	10
$\text{sign}(X_i - 30)$	$-$	$-$	$-$	$-$	$-$	$-$	$+$	$+$	$+$	$+$

Thus $W_o = (-)(7 + 5.5 + 5.5 + 3 + 2 + 1) + (+)(4 + 8 + 9 + 10) = -24 + 31 = 7$ which is less than the critical value of 10 (one-sided $H_A : \mu > 30$) and 8 (two-sided $H_A : \mu \neq 30$) from Ott and Longnecker (2001, Table 6, page 1099) at $\alpha = 0.05$ Thus, we accept H_o and conclude that the mean is **not** significantly different from $\theta_o = 30$

JMP provides the following results for the Wilcoxon signed-rank test (JMP uses $W_o/2$) by checking the Wilcoxon Signed Rank box shown in Figure 8.11 and using the data and the H_o specified above. Note that $H_o : \theta = 10$ is rejected by the (Wilcoxon) Signed-Rank test as in the Sign test, but accepted by the $t-$test ($p-$value = `Prob>t` = 0.1241)

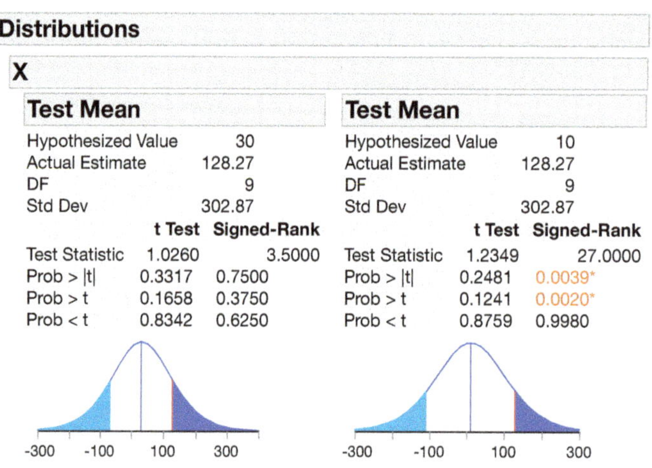

FIGURE 8.6. Wilcoxon Signed-Rank test vs $t-$tests

8.4.3 Mann-Whitney/Wilcoxon Rank-Sum Test

A nonparametric alternative to the two-sample $t-$test is the Mann-Whitney test, also called the Mann-Whitney-Wilcoxon test. This is equivalent to the Wilcoxon Rank Sum test.

Draw a random sample of size n_x from population X, and draw an independent random sample of size n_y from population Y. There are N observations in all, where $N = n_x + n_y$.

Rank all N observations. The sum $U = \sum \text{Rank}(Y_j)$ of the ranks for the Y sample is the Mann-Whitney statistic. The sum $W = \sum \text{Rank}(X_i)$ of the ranks for X sample is the Wilcoxon Ranked-Sum statistic. Since $U + W = N(N+1)/2$, these tests are equivalent in terms of $p-$values. JMP uses the Wilcoxon Ranked-Sum statistic.

If the two populations have the same continuous distribution: $H_o : F(x) = G(x)$ for all x, then under H_o the U and W statistics have mean and variance

$$\mu_U = n_y(N+1)/2 \qquad \mu_W = n_x(N+1)/2 \qquad \text{SE} = \sqrt{n_x n_y (N+1)/12}$$

The $Z-$approximation is $Z = \frac{U-\mu_U}{\text{SE}}$ or $Z = \frac{W-\mu_W}{\text{SE}} \approx N(0,1)$

We explore the Mann-Whitney/Wilcoxon rank sum test in the exercises

8.5 JMP Notes

The advantage of using software such as JMP (and Excel) is that they provide a common criterion called the $p-$value which we qualified in Table 8.1 That is, for any "test" that we will consider in this text, we can assess significance by comparing the $p-$value to 0.05

Here we look at the Placebo versus pain relief Drug dataset discussed in section 8.2.1 and shown again in Figure 8.7.

One-sample $t-$procedure

1. Open the `drugAB.jmp` dataset

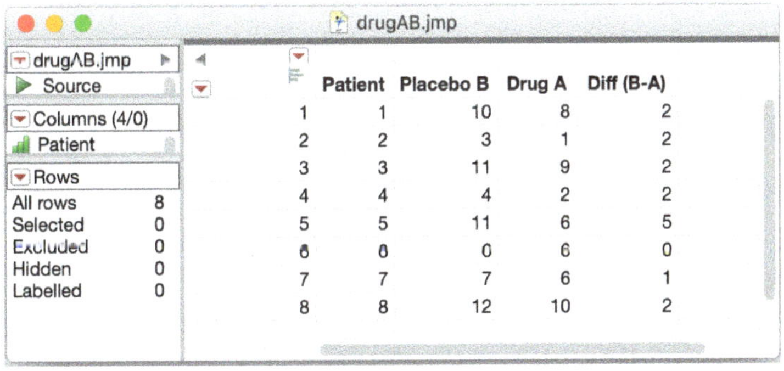

drugAB.jmp

	Patient	Placebo B	Drug A	Diff (B-A)
1	1	10	8	2
2	2	3	1	2
3	3	11	9	2
4	4	4	2	2
5	5	11	6	5
6	6	6	6	0
7	7	7	6	1
8	8	12	10	2

drugAB.jmp; Source; Columns (4/0); Patient; Rows; All rows 8; Selected 0; Excluded 0; Hidden 0; Labelled 0

FIGURE 8.7. drug dataset

2. Do Analyze → Distribution
3. Enter **Drug A** in the Y, Columns panel. Then click OK
4. In the report window do Distributions → Stack to make the display horizontal
5. JMP gives the 95% CI in the Summary Statistics report. (See Figure 8.10)

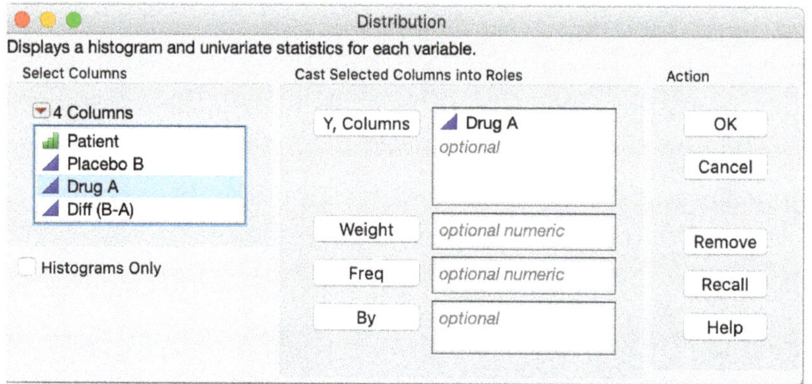

FIGURE 8.8. Distribution dialog window: drug dataset

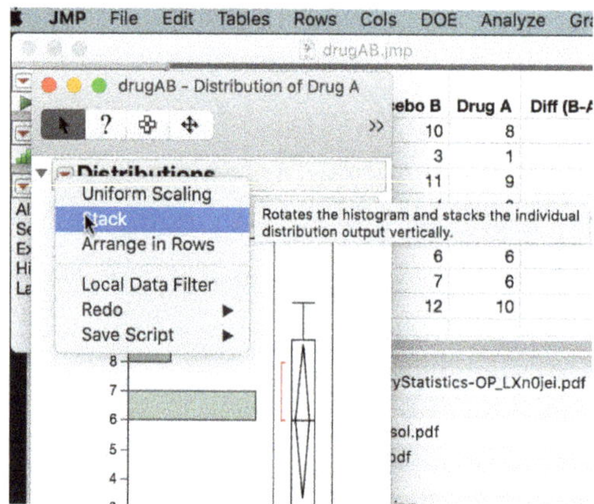

FIGURE 8.9. Distribution dialog window: drug dataset

NOTE: You can remove the Quantiles report by selecting Display Options in Figure 8.10 and "unchecking" Quantiles Similarly, you can remove the Summary Statistics report.

6. Now we want to "test" whether $\mu = 5$ (See Figure 8.10)

7. The result is shown in Figure 8.11 (After clicking OK in the Test Mean dialog popup window — do **NOT** enter a value in the True Standard Deviation box)

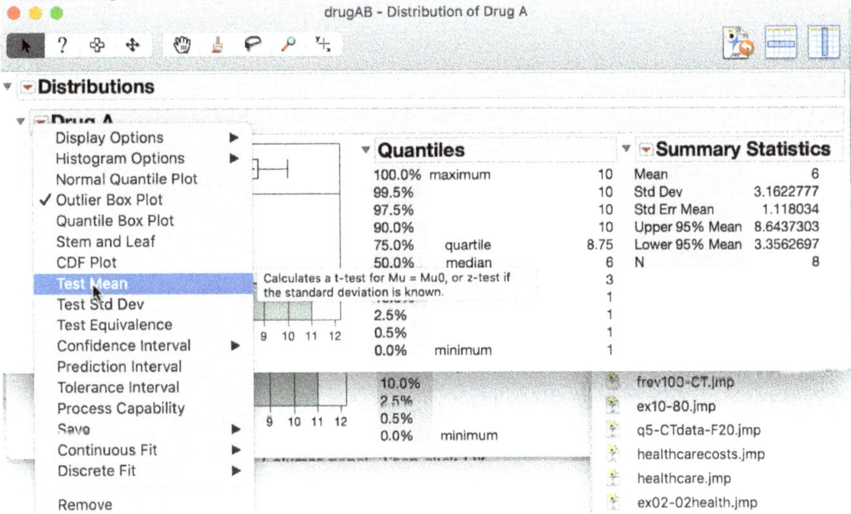

FIGURE 8.10. One sample $t-$test option: drug dataset

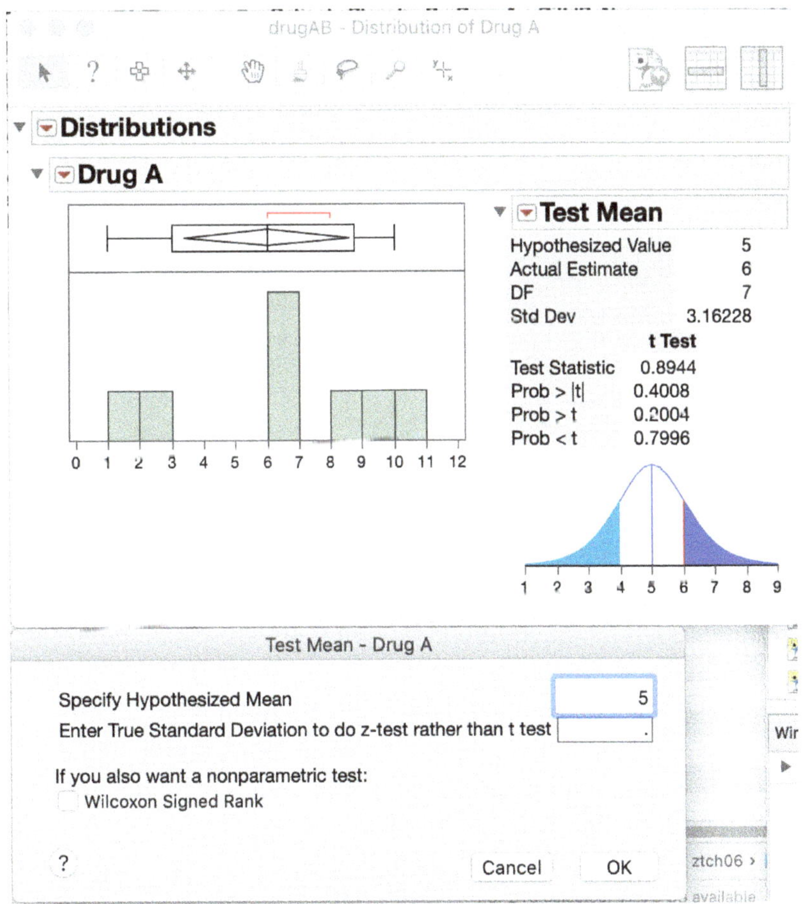

FIGURE 8.11. One sample $t-$test result: drug dataset

Paired $t-$procedure

1. Open the **`drugAB.jmp`** dataset

drugAB.jmp

	Patient	Placebo B	Drug A	Diff (B-A)
1	1	10	8	2
2	2	3	1	2
3	3	11	9	2
4	4	4	2	2
5	5	11	6	5
6	6	6	6	0
7	7	7	6	1
8	8	12	10	2

Columns (4/0): Patient

Rows: All rows 8, Selected 0, Excluded 0, Hidden 0, Labelled 0

FIGURE 8.12. drug dataset. Paired $t-$procedure

2. To conduct a **Paired $t-$procedure** we need to create a "Diff" column [shown as Diff(B-A) in Figure 8.12] For simple datasets like **`drugAB.jmp`** we "could" manually enter the differences. Fortunately, JMP can do this calculation for us.

3. Let us add a new column Column 5 and while it is highlighted, select **Cols → Formula** Result is shown in Figure 8.13.

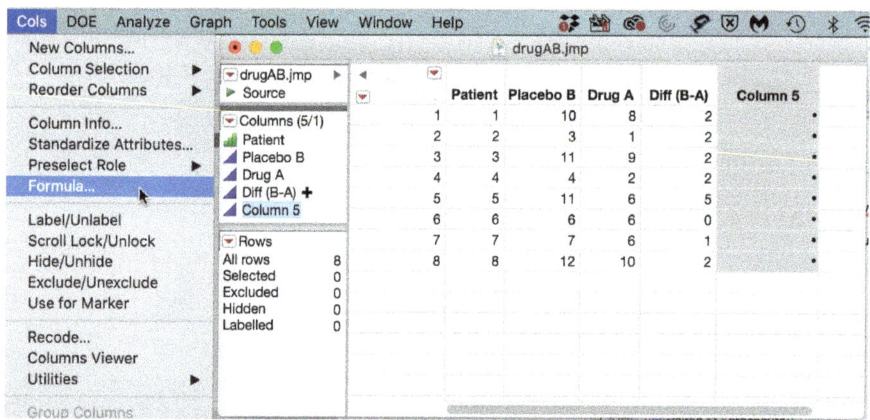

FIGURE 8.13. drug dataset: Diff Formula setup

4. In the **Formula** dialog window, enter Placebo B, then the "minus" sign, and then Drug A The result will look like Figure 8.14

   Click Apply to check the values entered in Column 5. If good, click **OK**

5. Now you can analyze "Diff" (Diff(B-A) or Column 5) using the One-sample $t-$ procedure

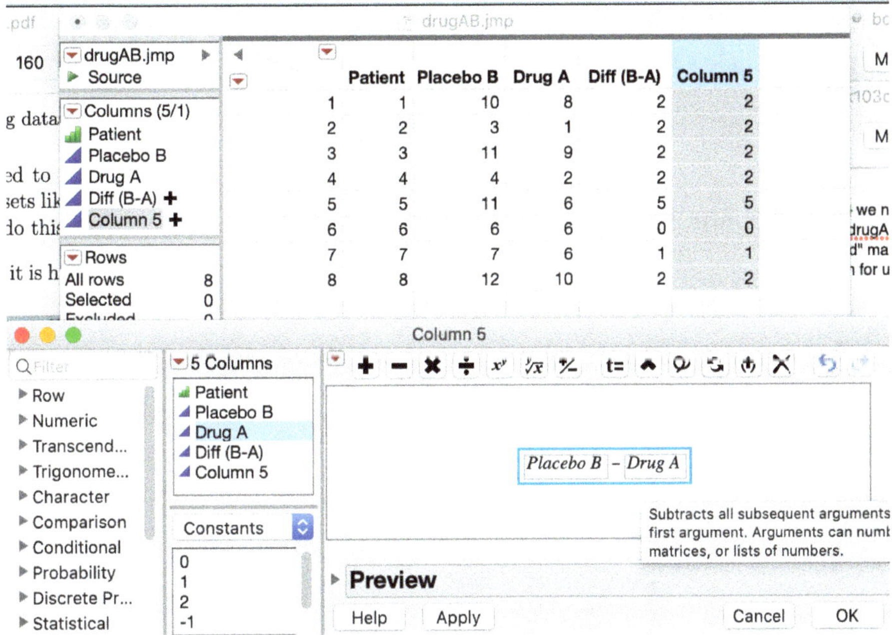

FIGURE 8.14. drug dataset: Diff Formula

Two-sample $t-$procedure

1. The **Two-sample $t-$procedure** requires a *two-level* $X = A$ Factor (categorical — Nominal or Ordinal data type in JMP) versus a Y response (quantitative)

2. If we pretend the `drugAB.jmp` dataset was administered to two independent groups of 8 patients, then we need to reformat by "stacking" the dataset

3. Do **Tables** → **Stack** Enter Placebo B and Drug A in the **Stack Columns** panel. Click **OK** (See Figure 8.15)

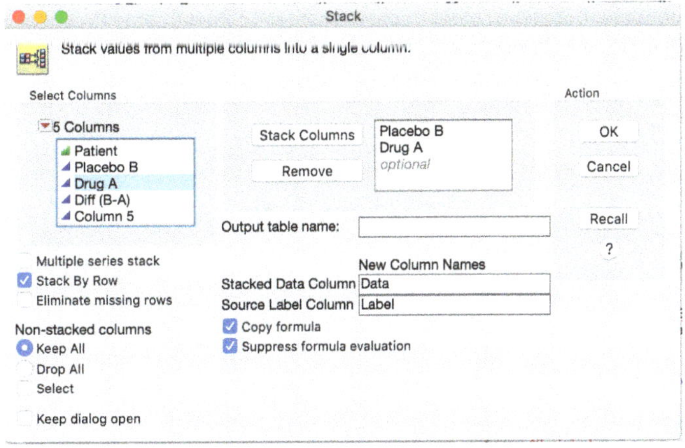

FIGURE 8.15. drug dataset: Stack panel

4. Now do **Analyze** → **Fit Y by X** Enter Data as the **Y, Response** and Label as the **X, Factor** Click **OK**

5. The result is shown in Figure 8.16.

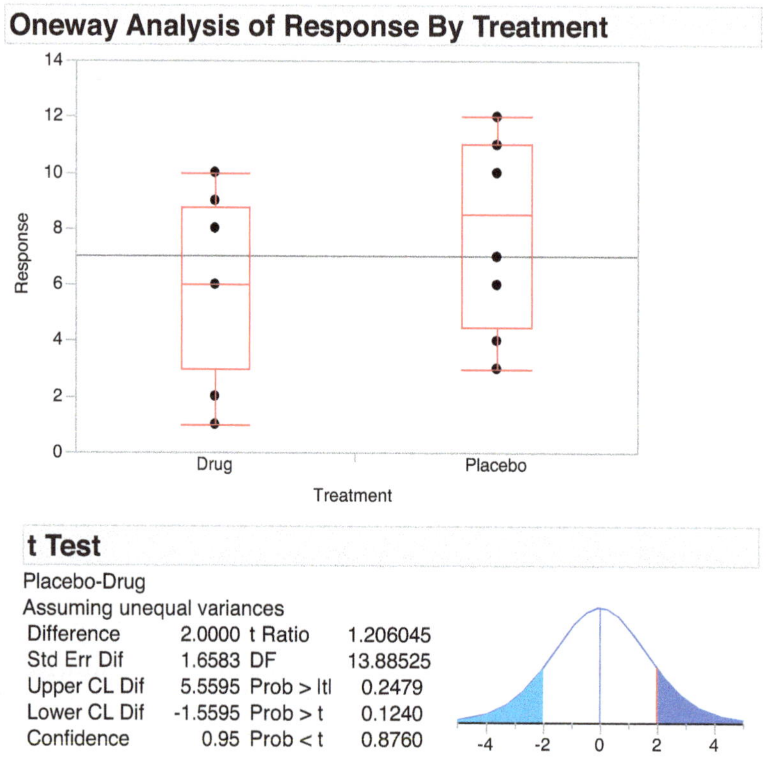

FIGURE 8.16. Independent $t-$Test

JMP $P-$values

For $t-$tests, JMP automatically lists three (3) $p-$values which depend on what alternative hypothesis (H_a) the researcher is interested in. The pre-experiment context is crucial

- `Prob > |t|` For change $H_a : \mu \neq \mu_o$ $H_a : \mu_D \neq 0$ $H_a : \mu_A \neq \mu_B$
- `Prob > t` For +ve change $H_a : \mu > \mu_o$ $H_a : \mu_D > 0$ $H_a : \mu_A > \mu_B$
- `Prob < t` For −ve change $H_a : \mu < \mu_o$ $H_a : \mu_D < 0$ $H_a : \mu_A < \mu_B$

Excel $t-$tests

Link: https://www.youtube.com/watch?v=q0ckcKsSPXU

Chapter 8 Exercises

8.1 BONELOSS Nursing mothers secrete calcium into their milk. Some calcium may come from their bones so mothers may lose bone mineral. The percent change in mineral content of the spines of 47 mothers was measured after 3 months of breastfeeding. This was recorded as Change in the dataset. Negative values means bone loss occurred.

a) Open the **boneloss** dataset and create a histogram and stemplot.

b) Conduct the $t-$test for $H_o : \mu_{\text{Change}} = 0$

c) What is the appropriate P value from the $t-$test report? Explain why.

d) What is your conclusion. Be specific.

8.2 The HIGH and LOW temperature forecasts for 14 U.S. Cities were recorded in early April. Assume that the high and low temperature forecasts are normally distributed.

a) Explain why testing $H_o : \mu_L = \mu_H$ would **not** be sensible for this data.

b) Explain why a matched pairs procedure should be used to analyze this data.

c) A 95% confidence interval for μ_D (where μ_D = mean of differences HIGH − LOW) is 15.38094 to 24.61906 Using this information, compute the following

i) The sample mean $\overline{d}$ of the differences.

ii) The standard deviation s_d of the differences. Round to 1 decimal place.

8.3 Rents for one-bedroom and two-bedroom apartments are displayed in Figure 8.17

a) Suppose you would like to rent a two-bedroom apartment. What feature of the boxplot display is particularly relevant to your preference . . . or wallet?

b) The standard error (SE) of $(x_1 - x_2)$ is 38.5 (to 1 dp).

i) Hence, what is the value of the test statistic to for testing $Ho : \mu_1 = \mu_2$

ii) The $P-$value for HT above is prob $> |t| = 0.0703$ What is your conclusion ?

c) Does your conclusion change if $H_A : \mu_1 < \mu_2$ Explain.

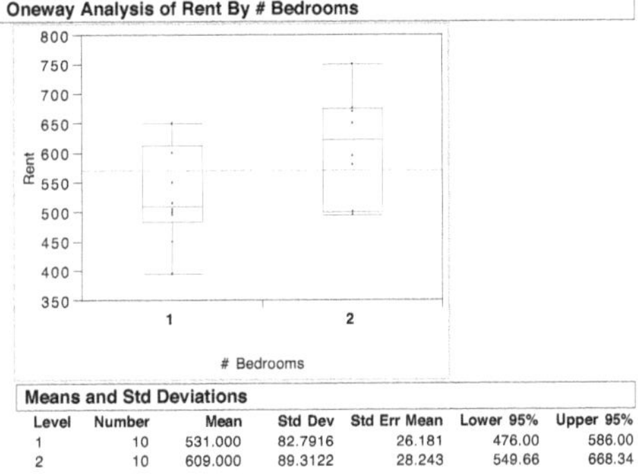

Means and Std Deviations

Level	Number	Mean	Std Dev	Std Err Mean	Lower 95%	Upper 95%
1	10	531.000	82.7916	26.181	476.00	586.00
2	10	609.000	89.3122	28.243	549.66	668.34

FIGURE 8.17. Apartment Rents

8.4 CHOLESTEROL 18 patients currently on a higher dosage of a cholesterol drug were switched to the Lower dosage for 3 months. Here, Diff = Low - High

a) Perform a paired $t-$test with $H_o : \mu_{\text{diff}} = 0$

b) State your conclusion.

c) Your conclusion applies to *all* patients who could be switched to a Lower dosage Should this "statistical" result dictate standard medical practice? Explain.

8.5 CHOLESTEROL The problem with the analysis in Exercise 8.4 is that it is known that *on average* patients on the Lower dosage will experience an increase in cholesterol levels. Physicians only recommend switching patients if their cholesterol increases by 5 or less points This is an example of an "equivalency" test.

a) Perform a new paired $t-$test with $H_o : \mu_{\text{diff}} = 5$

b) What is the appropriate alternative hypothesis? Explain why.

c) What is your conclusion regarding the *equivalency* of the Lower dosage ?

8.6 It is suspected that the lifetime (in weeks) is different for two brands of batteries. Five randomly selected batteries of each brand (X and Y, say) were tested and the average lifetimes were: $\overline{x} = 79.4$ and $\overline{y} = 100.4$ with variances: $s_x^2 = 40$ and $s_y^2 = 64$

a) Construct a 95% CI for the mean lifetime of battery brand X.

b) Compute the standard error SE for this two-sample $t-$test

c) We will use a *conservative* $t-$test with df = 4

i) Explain what this means.

ii) Hence, carry out the appropriate HT and state your conclusion

d) Is it possible that one Brand X battery lasted as long as the mean of brand Y? Explain your reasoning. [HINT: Check s_x^2]

8.7 WILCOXON Verify the result of Figure 8.6

JMP: Test Mean dialog Enter 10 as the Hypothesized Mean and Check the Wilcoxon Signed Rank box

8.8 MANNWHITNEY Show that the Mann-Whitney test is significant for $H_o : \mu_X = \mu_Y$ whereas the 2-sample $t-$test is **not** significant.

JMP: **Fit Y by X** and enter Data as Y and Group as X

Select t Test and then select Nonparametric → Wilcoxon Test

9
Data Analytics

We have now covered some of the standard Statistical methods that are commonly used in what has become known as **Data Analytics**. This can be described as a *decision-orientated* process in the sense that some type of definitive action (decision) "may" be taken based on the results and conclusions obtained by the data analytics team.

9.1 Overview

An example of data analytics is our FICO score. FICO is a credit scoring model originally developed by the **F**air **I**saac **CO**rporation company which uses *predictive analytics* to assess our credit worthiness. That is, our FICO score provides the lender with an indicator of our reliability to repay a loan such as a house mortgage, car loan, or credit card balances.

The COVID-19 example shown in Figure 1.4 illustrates the real-time role of data analytics. Even with limited data, the trend showed that a definitive decision needed to madc to shut down nonessential businesses (despite the economic cost) and implement social distancing and masking requirements to help prevent the spread of COVID-19.

Data analytics is a multidisciplinary field and problem-specific. It is used extensively in business and finance, marketing, online systems, information security, and software services.

So how much data is created every day? The 2023 estimate by Earthweb[1] is:

3.5 quintillion bytes every day = 3.5×10^{18} bytes = 3.5 billion GB (gigabytes)

[1] https://earthweb.com/how-much-data-is-created-every-day

We should note that not all data created is "stored." For example, when you make a phone call, your conversation is actually digitalized (in real-time $\Rightarrow$ data created), transmitted, and then "unpacked" for your recipient to hear. When your call ends, that data is gone, but will be counted as data created. Your phone carrier will record the date, time, and duration of your phone call for billing purposes, but not your actual conversation.

Key Point: Data analytics requires extensive computation to work with **big data**.

The point is that data analytics can require extensive computation to work with **big data**. The term *advanced data analytics* is used to describe the more technical aspects of data analytics such as the use of **machine learning** (see Kelleher, et al. 2020), logistic regression, decision trees, linear and multiple regression models, time series analysis, and classification to do predictive modeling. Other techniques include cluster analysis, principal component analysis, segmentation profile and association analysis (see Myers 2019).

The above topics are beyond the scope of this book, but show that data analytics can require advanced statistical methods to solve certain problems. In the following sections we consider some examples of data analytics.

9.2 Body Roundness Index

Obesity is known to increase the risk of adverse health effects such as type 2 diabetes, cardiovascular disease and atherosclerosis[2] (see Brochu et al. 2000) In section 2.6 we defined Body Mass Index (BMI).

$$\text{BMI} = \frac{\text{Weight(kg)}}{[\text{Height(meters)}]^2} = 703 * \frac{\text{Weight(lbs)}}{[\text{Height(in)}]^2}$$

A person's BMI *number* is widely used to classify their weight status as follows:

TABLE 9.1. BMI Classifications

Classification	BMI
Underweight	< 18.5
Healthy	$18.5 - 24.9$
Overweight	$25 - 29.9$
Obese	$30 - 39.9$
Extremely Obese	≥ 40

[2]This relates to the narrowing of the artery wall due to the buildup of plaque and can result in coronary artery disease, stroke, peripheral artery disease, or kidney problems. See https://en.wikipedia.org/wiki/Atherosclerosis

The problem is that BMI does not necessarily give a good proxy estimate of **body fat** percentage. Furthermore, BMI can differ based solely on your gender! Thus, BMI has often been criticized for misclassifying individuals at risk for obesity related comorbidities.

Key Point: Dig deeper *What is body fat?*

Total body fat mostly includes visceral adiposity[3] ("beer belly") and subcutaneous fat that lies below the skin in the hypodermis region (such as the arms and buttocks). Subcutaneous fat is considered to pose less of a health risk to obesity-related pathologies compared to visceral fat.

Although imaging methods such as dual energy X-ray absorptiometry (DXA) and magnetic resonance imaging can directly quantify adiposity, most individuals are not imaged, so an easily obtained anthropometric measure is desirable.

Thomas et al. (2013) derived a body roundness index (BRI) to quantify the individual body shape using the geometry of an ellipse. In 2-dimensions (2D) Figure 9.1 depicts BMI (rectangle) and BRI (red oval) that encapsulate the human body in terms of height and width. In 3D these correspond to a cylinder (BMI) and ellipsoid (BRI) (like a football or rugby ball).

Body roundness improved predictions of %body fat and %visceral fat compared to the traditional metrics BMI, Waist and Hip circumference.

Conclusion: Learn the science involved in the problem.

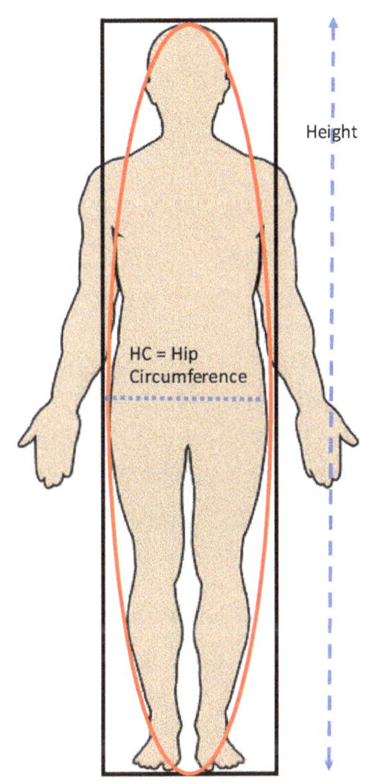

FIGURE 9.1. BMI versus BRI

9.3 Business Intelligence

One of the earliest adopters of data analytics was the financial sector. Data analytics has an important role in the banking and finance industries, that is used to predict market trends and assess risk. It is now called **Business Intelligence** (BI).

The following links provide the basics of data analytics in the context of BI

https://www.youtube.com/watch?v=yZvFH7B6gKI

https://www.investopedia.com/terms/b/business-intelligence-bi.asp

[3] https://en.wikipedia.org/wiki/Adipose_tissue

9.4 Precipitation

Global warming and climate change have made an impact on our lives. The last three decades has been successively warmer at the Earths surface than any preceding decade. Precipitation is a major component of the hydrologic cycle and is essential for life on Earth. Overland, it is a primary source for all freshwater used for general water consumption. It contributes to the maintenance of soil moisture and replenishes natural underground water reservoirs while at the same time surface runoffs add to streams and ultimately flow out into oceans.

As temperatures rise and the Earths surface warms, characteristics of precipitation such as amounts, duration, and frequency are directly influenced and altered. However, in areas that have a lack of precipitation, this can have devastating effects such as severe drought which can lead to further issues like increased risk of wildfires and heatwaves.

Changes in the major ocean currents caused by global warming is also affecting precipitation patterns. It is known that the variability of precipitation in the tropics is highly influenced by the cyclic variations of sea surface temperatures (SST) across the equatorial Pacific caused by the El Nino Southern Oscillation (ENSO) every two to seven years. However, since 1975, there has been a shift in temperatures to warmer conditions which have caused the ENSO events to have become more intense, frequent, and persistent.

The state of New Jersey is located in northern United States and usually experiences a wide range of temperatures throughout the year with warm summers during the months from June to September and cold wet and snowy winters during the months of December to February. Typically, there is more rainfall during the cooler seasons than in the warmer.

The analysis of NJ precipitation from McDougall and Ahmed (2017, modified) is shown in Figure 9.2

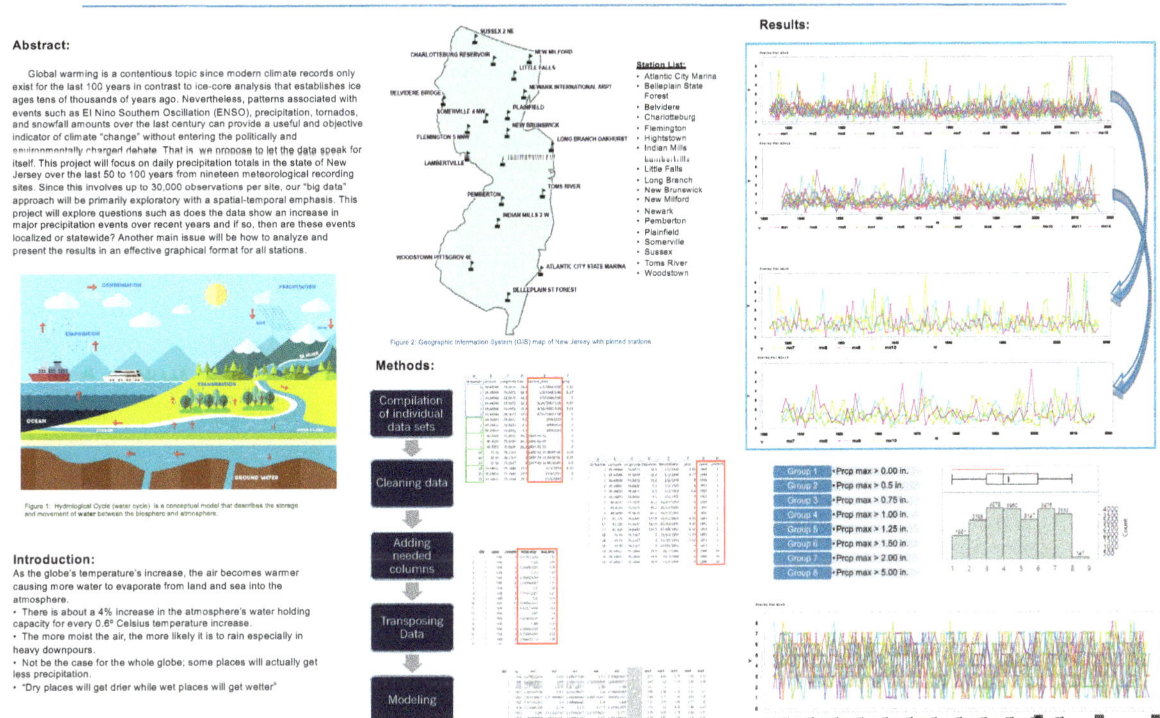

FIGURE 9.2. NJ Precipitation Analysis

9.5 Multiple Linear Regression

The simple linear regression (SLR) model discussed in chapter 5 can be extended to include additional predictors. The *multiple linear regression* (MLR) model with p predictors is:

$$y_i = \beta_0 + \beta_1 x_{1i} + \ldots + \beta_p x_{pi} + \epsilon_i$$

where β_0 is the intercept (the expected value of Y when all the predictors $X_j = 0$), and the other β_j's are sometimes called the *partial* slopes. That is, β_j is the average change in Y when X_j is increased by one "unit" and **all** the other predictors are held constant. As in the SLR case, we assume:

1. The MLR model specification is correct
2. $E[\epsilon_i] = 0$ and $Var[\epsilon_i] = \sigma^2$

3. The ϵ_i are independent and do not depend on the X_j's

4. For classical inference purposes $\epsilon_i \sim N(0, \sigma^2)$

The *linear* form of the MLR model is not actually a major restriction since it allows a wide variety of "models" to be investigated. For example:

Polynomial Regression $\widehat{Y} = \beta_0 + \beta_1 X + \beta_2 X^2 + \ldots + \beta_p X^p$ [i.e., $X_j = X^j$]

Interactions Let $X_3 = X_1 * X_2$ Then $\widehat{Y} = \beta_0 + \beta_1 X_1 + \beta_2 X_2 + \beta_3 X_3$
allows the interaction effect $X_1 X_2$ to be analyzed

ANOVA Factors Let $X_1 = 0$ if `GENDER` = Male, and $X_1 = 1$ otherwise. Then
$\widehat{Y} = \beta_0 + \beta_1 X_1 + \beta_2 X_2 + \ldots$ includes the `GENDER` factor

9.6 The General Linear Model (GLM)

The terminology MLR is usually reserved for the situation where all the X_j predictors are quantitative. However, we have seen that the use of *dummy variables*, $X_j = 0$ or 1, enable ANOVA (categorical) factors to be represented as a MLR model[4]. Thus, when ANOVA factors are included, the MLR model is referred to as a *General Linear Model* (GLM).

Dummy Variables To represent a categorical factor with k levels, we need $k - 1$ *dummy variables.* For example, if `TEMP` has 3 levels, `LOW, MED, HIGH` we define:

$$\begin{array}{llll} X_1 = 1 & \text{if TEMP} = \text{LOW} & \text{and} & X_1 = 0 \quad \text{otherwise} \\ X_2 = 1 & \text{if TEMP} = \text{MED} & \text{and} & X_2 = 0 \quad \text{otherwise} \end{array}$$

Thus, if both $X_1, X_2 = 0$ then `TEMP` must be `HIGH`[5]

Example 9.1 Let Y be the Salary, G the gender of the employee, and X be the number of months of employment. The ANCOVA model with interaction is: $Y = G + X + GX$

The MLR representation of this model can be obtained by setting $X_1 = 1$ if $G =$ Male, and 0 if $G =$Female (or the reverse), $X_2 =$ Months employed, and $X_3 = X_1 * X_2$ as the GX interaction. Thus

$$Y = \beta_0 + \beta_1 X_1 + \beta_2 X_2 + \beta_3 X_3 + \epsilon$$

which actually provides two separate SLR models for Males and Females:

$$\begin{aligned} Y_M &= (\beta_0 + \beta_1) + (\beta_2 + \beta_3) X_2 + \epsilon \quad \text{since} \quad X_1 = 1 \\ Y_F &= (\beta_0) + (\beta_2) X_2 + \epsilon \quad \text{since} \quad X_1 = 0 \end{aligned}$$

---

[4]All software packages actually use the MLR representation to compute ANOVA designs

[5]Note that the condition $X_1 = 1, X_2 = 1$ is **not** used since this would imply `TEMP = LOW` "and" `MED`

Hence, if both β_1 and β_3 are greater than zero, the MLR would show that the starting salary ($X_2 = 0$) of Males is higher than Females, and the salary of Males increases at a faster rate than Females! This is illustrated in the schematic below.

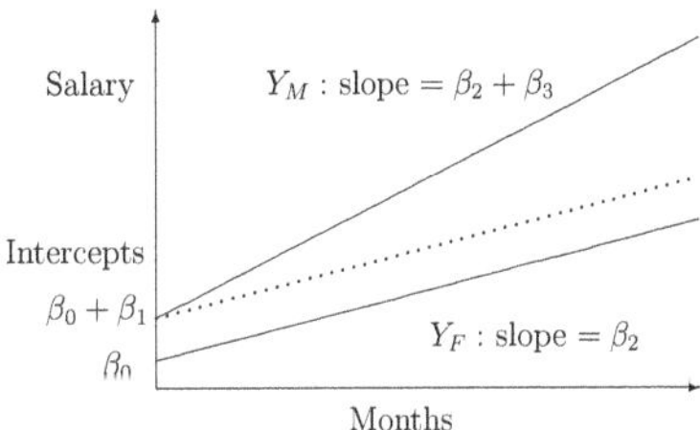

FIGURE 9.3. Hypothetical SLR example for Salary vs Months by Gender

9.7 Fitting the MLR

In the SLR case ($p = 1$) the LS fit $\widehat{Y} = \beta_0 + \beta_1 X_1$ is a straight line. For $p = 2$ we have $\widehat{Y} = \beta_0 + \beta_1 X_1 + \beta_2 X_2$ which geometrically is a plane in 3D (imagine a sheet of paper held up in the air in some orientation). Clearly, for $p > 2$ we cannot visualize the "hyperplane" completely. This means we will be much more reliant on numerical and residual diagnostics to assess the fitted MLR.

Least Squares Estimates The LS parameter estimates $\hat{\beta}_j$ minimize SSE $= \sum_i e_i^2$ where the residual $e_i = y_i - \hat{y}_i$ (observed - fitted value).

The SLR diagnostics discussed in chapter 5 can therefore be employed for assessing the fitted MLR. For example, the quality of fit can be measured by RSquare:

$$\text{RSquare} = 1 - (\text{SSE} / \text{SST}) \qquad \text{where} \quad \text{SST} = \sum (y_i - \overline{y})^2$$

Similarly, we can visually examine the residual versus fitted plot, compute studentized residuals (MLR refitted leaving out the ith observation) and Cook's D to assess influential points, and use Q-Q plots and the Shapiro-Wilk test to assess normality.

We will rely on software such as JMP to do the computations. So let's consider an example where the scores (out of 40) from the first two assignments were used to predict the student's midterm (out of 100) performance. The data is shown in Table 9.2

The **Fit Model** platform in JMP was used to fit the MLR prediction model: $\widehat{Y} = \beta_0 + \beta_1 X_1 + \beta_2 X_2$ where Y = MT score and X_1, X_2 are the assignment A1, A2 scores. Then the **Save Columns** $\rightarrow$ option from the MLR report was used to save the residuals, studentized residuals, and Cook's D. For Student #1 Cook's D = 1.743 > 1 and indicates that this observation has high influence. The MLR report is shown in Figure 9.4

TABLE 9.2. Midterm Dataset and MLR diagnostics

	A1 40	A2 40	MT 100	Residual	Student	Cook's D
1	24	19	69	5.503	1.142	1.743
2	28	32	68	6.407	0.752	0.113
3	31	38	57	-5.899	-0.818	0.277
4	36	38	72	-0.363	-0.035	0.000
5	34	32	66	-6.950	-0.683	0.020
6	38	36	59	-18.606	-1.818	0.122
7	40	40	95	16.524	1.620	0.103
8	39	39	97	19.688	1.908	0.112
9	36	34	74	-1.278	-0.124	0.001
10	38	38	69	-7.148	-0.687	0.012
11	36	37	60	-13.091	-1.258	0.039
12	38	38	75	-1.148	-0.110	0.000
13	39	39	87	9.688	0.939	0.027
14	40	39	73	-6.205	-0.609	0.015
15	38	37	80	3.123	0.301	0.002
16	39	38	85	6.959	0.676	0.015
17	40	39	72	-7.205	-0.708	0.020

Student #1 is represented by the bold point in the graphs of Figure 9.4 and does not appear to be at all unusual in the **Actual** or **Residual** plots. Only in the 3D **Surface** plot do we see that the point is substantially removed from the other observations and the left corner of the green plane is close to the Student #1 point. We are actually looking at the *underside* of the plane so we can see the points below. In JMP, the surface plot is interactive so you can move it around to get a suitable perspective.

Interpretion of the MLR We essentially use the same approach as in the SLR case:

Summary of Fit

RSquare = 0.248 so the MLR fit explains less than 25% of the variation in the MT scores. The quality-of-fit of the MLR model is rather poor.

RMSE = 10.8 (residual standard deviation) and all the Residual are within $\pm 2 * 10.8$

Analysis of Variance

The overall $F-$test is not significant Prob >F = 0.1363 so none of the "Factors" X_1, X_2 in the MLR model appear to be useful for predicting MT score.

Lack of Fit

Not really useful here since only 69 and 72 are repeated MT scores. However, there is weak evidence Prob >F = 0.0796 that the MLR model is not "adequate"

Parameter Estimates

This differs from the SLR case since the estimates for X_1 = A1 and X_2 = A2 are partial slopes. The significance of each X_j is also assessed separately.

Intercept: β_0 = 31.92 is the predicted MT score if a student didn't do the first two assignments. This actually makes some sense in the context of this dataset.

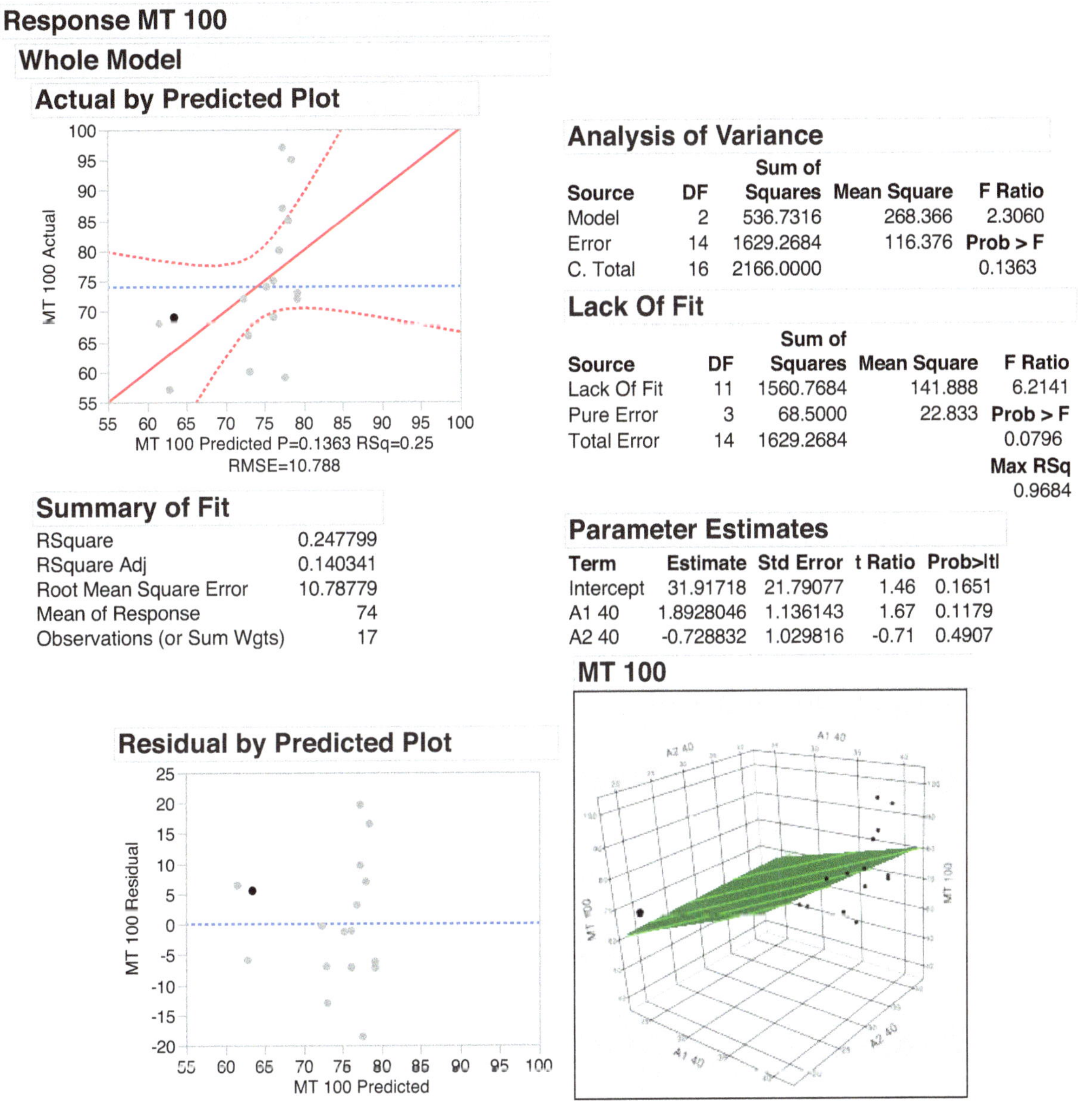

Response MT 100

Whole Model

Actual by Predicted Plot

Analysis of Variance

Source	DF	Sum of Squares	Mean Square	F Ratio
Model	2	536.7316	268.366	2.3060
Error	14	1629.2684	116.376	**Prob > F**
C. Total	16	2166.0000		0.1363

Lack Of Fit

Source	DF	Sum of Squares	Mean Square	F Ratio
Lack Of Fit	11	1560.7684	141.888	6.2141
Pure Error	3	68.5000	22.833	**Prob > F**
Total Error	14	1629.2684		0.0796
				Max RSq
				0.9684

Summary of Fit

RSquare	0.247799
RSquare Adj	0.140341
Root Mean Square Error	10.78779
Mean of Response	74
Observations (or Sum Wgts)	17

Parameter Estimates

Term	Estimate	Std Error	t Ratio	Prob>|t|
Intercept	31.91718	21.79077	1.46	0.1651
A1 40	1.8928046	1.136143	1.67	0.1179
A2 40	-0.728832	1.029816	-0.71	0.4907

Residual by Predicted Plot

FIGURE 9.4. MLR output for Midterm dataset

X_1: $\beta_1 = 1.89$ implies that doing the first assignment increases the average MT score by 1.89 points for every 1 point increase in the A1 score.

X_2: $\beta_2 = -0.73$ implies doing the second assignment *decreases* the MT score !? Looking at A2 scores we see that all but one were 30+ and a large number were above 37 However, this "expected" performance was not reflected in the student's MT scores.

RSquare Adj For a particular MLR fitted model, RSquare provides a measure of the quality-of-fit for that model. The problem is that RSquare will *always* increase by simply adding more X_j predictors (even if the added X_j consists of completely random values). Indeed, if we have $p = n - 1$ predictors (i.e., $p = n$ parameters β_j) we get RSquare $= 1$

KP Model complexity p needs to be compatible with the *information* available n

$$\text{RSquareAdj} = 1 - \frac{(n-1)\,\text{SSE}}{(n-p)\,\text{SST}} = 1 - \frac{\text{MSE}}{\text{MST}}$$

where MST $= \text{SST}/(n-1)$ is the variance of the responses Y_i Thus, RSquare Adj includes a *penalty* term $(n-1)/(n-p) > 1$ which trades-off the reduction in SSE when more predictors X_j are added. In fact, RSquare Adj can become *negative* when p is large relative to n

The interpretation of RSquare Adj is the same as with RSquare, but the biggest advantage is that RSquare Adj allows MLR models with different subsets of predictors to be compared in terms of quality-of-fit. For the Midterm example, we could compare the two separate SLRs with the MLR above[6].

Inference for MLR The Parameter Estimates panel in Figure 9.4 also provides $t-$tests for the hypothesis: $H_o : \beta_j = 0$ versus $H_A : \beta_j \neq 0$ for each parameter:

$$\textbf{t Ratio} = \frac{\textbf{Estimate}}{\textbf{Std Error}} \sim t_{n-p} \quad \text{under } H_o$$

Thus, **Prob > t** is the $P-$value which gives the significance of each parameter in the MLR model. Equivalently, it assesses the usefulness of X_j for predicting Y *assuming* all the other predictors are already in the MLR model. That is, does X_j contribute significantly to predicting Y beyond that already provided by the other predictors. For the Midterm data, neither assignment score appears to be useful for predicting the MT score.

Key Point: A nonsignificant X_j in one MLR model may be significant in a different MLR model

That is, we cannot conclude that "both" X_1 and X_2 are not useful predictors of Y in a SLR. Indeed X_1 by itself is a significant predictor of Y (even though RSquare is still poor). This means choosing a "good" subset of predictors cannot be based simply on the significance of the predictors currently included in the MLR model.

[6]Excluding Student #1 the SLR with A1 has RSquare Adj $= 0.25$ versus 0.20 for the MLR (without Student #1)

Chapter 9 Exercises

9.1 Fix BMI = 25.

Find three (3) different Weight and Height combinations that satisfy this BMI

9.2 RANDU is a linear congruential pseudorandom number generator (LCG) of the Park-Miller type, which was used primarily in the 1960s and 1970s for simulations. In the sample dataset 1500 observations were generated using Randu and split into three columns.

a) Show that the x, y, z all appear to follow a U(0,1) distribution

b) Show that any scatterplot also shows a random pattern

c) In JMP select **Graph → Scatterplot 3D** Enter x, y, z as Y

Rotate to see that the points fall into two-dimensional planes.

9.3 SALARY The dataset contains Gender, Months (on job), and Salary in $1000s Here we use **Analyze → Fit Model** for analyzing Example 9.1

a) Fit the ANCOVA model.

Do **Analyze → Fit Model**
In the popup window, enter Salary as Y and click on the "Keep dialog open" box
Click on Gender and Months in the top left column list. Then click on **Add**
Again click on Gender in the top left column list, but click on Months in **Add**
Then click on **Cross** and **Emphasis→ Minimal Report** Then **Run**

b) Match the parameter estimates to Figure 9.3

9.4 COLLEGE debt presents a financial burden for many students. Here, we consider predicting Y = Average student debt from n=27 colleges using the predictor measures:

Admit	=	admitted proportion
Grad4Rate	=	proportion that graduate in 4 years
InCostRate	=	average tuition cost for in-state students
InCostAid	=	average aid (scholarships, assistantships, etc.) for in-state students

a) Fit the MLR using all the predictors.

JMP Notes: **Analyze → Fit Model**
In the popup window, click on "Keep dialog open" box then enter AveDebt as Y
Click on Admit in the top left column list. Then click on **Add**
Do the same for Grad4Rate InCost, InCostAid. Then click **Run**

b) State your conclusion.

9.5 COLLEGE Find the best MLR model based on the largest RSquare Adj

That is, which combination of the predictor measures gives the largest **RSquare Adj** value

10
Analysis of Variance (ANOVA)

The two-sample $t-$test procedure is used when the *same* quantitative response Y is measured on *two* independent samples. For example, Y = `SAT` math score versus A = `GENDER` where the objective is to assess whether there is (statistical) evidence of a gender bias with respect to the math portion of the SAT examination. In this case, we could test

$$H_o : \mu_F = \mu_M \quad \text{versus} \quad H_A : \mu_F \neq \mu_M$$

using the *pooled* two-sample $t-$test since it would be reasonable to assume that the *variation* in male and female `SAT` math scores are equal, even if a bias does exist.

What if we have $k > 2$ levels associated with the group factor A ? Let Y = `YIELD` of corn from homogeneous one-acre fields that were applied with fertilizer containing different levels of A = `NITROGEN` e.g., 0% (no fertilizer = "control" =1) , 5% (low=2), 10% (medium=3), 15% (high=4). Here $k = 4$, so intuitively the hypothesis to test is

$$H_o : \mu_1 = \mu_2 = \mu_3 = \mu_4 \quad \text{versus} \quad H_A : \textit{at least two levels differ}$$

that we might try to test using the the two-sample $t-$test on all 6 pairwise comparisons:

$$H_o : \mu_1 = \mu_2 \quad H_o : \mu_1 = \mu_3 \quad H_o : \mu_1 = \mu_4$$

$$H_o : \mu_2 = \mu_3 \quad H_o : \mu_2 = \mu_4 \quad H_o : \mu_3 = \mu_4$$

The problem is that these "tests" are **not** the same. $H_o : \mu_1 = \mu_2 = \mu_3 = \mu_4$ would employ *all* the data whereas the individual $t-$tests will only use the data from each level.

Key Point: We need an overall method for analyzing $k > 2$ groups.

10.1 One-way ANOVA

It may seem somewhat strange that we are considering a method based on the analysis of *"Variance"* to test a hypothesis about "means" since we know the sample variance provides no information about the sample mean! Indeed, if we were to reject $H_o : \mu_1 = \mu_2 = \mu_3 = \mu_4$ this does *not* tell us which nitrogen level(s) provided the greatest corn yield.

Key Point: The method of ANOVA is a two stage conditional analysis:

1. An overall ANOVA "test" of $H_o : \mu_1 = \mu_2 = \ldots = \mu_k$ is performed
2. Only if H_o is **rejected** do we perform a *post hoc* analysis of the "means"

Let us consider the corn example with `YIELD` data collected from 20 one-acre fields with five randomly selected plots allocated for each of the four fertilizer `NITROGEN` levels.

TABLE 10.1. One-way ANOVA Example

	`NITROGEN` Level $= i$			
Plot $= j$	0%	5%	10%	15%
1	27	35	44	31
2	30	38	39	37
3	25	40	40	35
4	27	39	45	33
5	26	43	42	34
Mean $\overline{y}_i$	27	39	42	34
Variance s_i^2	3.5	8.5	6.5	5.0

The sample means suggest a low 5% or medium 10% levels of `NITROGEN` in the fertilizer increased the `YIELD` of corn versus the Control 0%, but too much Nitrogen appears to have an adverse effect. The question is whether the study data provide sufficient evidence to conclude that the true mean yields μ_i $i = 1, 2, 3, 4$ are not all equal. If there is *insufficient* evidence then we accept H_o which corresponds to the conclusion that $\mu_i = \mu$ and thus the "Treatment" `NITROGEN` has no effect on the "Response" `YIELD`. That is, the group sample means $\overline{y}_i$ do not differ significantly from the overall sample average $\overline{y} = 35.5$

ANOVA Model The population one-way ANOVA model can be written as:

$$\texttt{DATA} = \texttt{MODEL} + \texttt{ERROR} \quad \Rightarrow \quad y_{ij} = \mu_i + \epsilon_{ij} \quad \Leftrightarrow \quad y_{ij} = \mu + \tau_i + \epsilon_{ij}$$

where y_{ij} = Response (`YIELD`) at the i'th Treatment (`NITROGEN`) level for the j'th Subject (`PLOT`); μ_i = i'th Treatment level mean; μ = overall Response mean; and τ_i = i'th Treatment level Effect $= \mu_i - \mu$ The fitted ANOVA model is therefore

$$y_{ij} = \overline{y}_i + (y_{ij} - \overline{y}_i) \quad \text{or} \quad y_{ij} = \overline{y} + (\overline{y}_i - \overline{y}) + (y_{ij} - \overline{y}_i)$$

where the fitted Treatment effect model (second equation) can be written as

$$(y_{ij} - \overline{y}) = (\overline{y}_i - \overline{y}) + (y_{ij} - \overline{y}_i)$$

We see that squaring and summing the left side of the above over all the observed y_{ij} would give $\sum_{ij}(y_{ij} - \overline{y})^2$ which is simply the numerator of the sample variance of the observed data $\{ y_{ij} \}$ — hence the terminology *Analysis of Variance.* It can be shown[1] that

$$\sum_{ij}(y_{ij} - \overline{y})^2 = \sum_{ij}(\overline{y}_i - \overline{y})^2 + \sum_{ij}(y_{ij} - \overline{y}_i)^2$$

which we will denote as

$$\text{SST} = \text{SSA} + \text{SSE}$$

where SS = Sum-of-Squares, so SST = Total SS , SSA = Treatment SS for factor A, and SSE = Error SS. Since SSA represents the differences between the group sample means $\overline{y}_i$ relative to the overall sample mean $\overline{y}$ it is also referred to the *Between Group SS*. Similarly, SSE is referred to as the *Within Group SS.*

It follows that if the variation *Between* the Treatment means is large compared to the *Within* group variation, we should reject $H_o : \mu_1 = \mu_2 = \ldots = \mu_k$ But, "variation" needs to be computed as an *average* measure i.e., $s_y^2 = \text{SST}/N - 1$ where $N = 20$ for the corn data.

Mean Sum-of-Squares MS = SS / df
MS is the "average" SS which depends on the df (degrees-of-freedom) Using the same argument for the sample variance s_y^2 — if we know $N - 1$ of the observed data $\{ y_{ij} \}$ and the sample mean $\overline{y}$ then we know the missing y_{ij} For SSA, we have k group sample means and the overall sample mean. Hence $\text{df}_A = k - 1$ Similarly, $\text{df}_E = N - k$ Thus, the one-way ANOVA test for $H_o : \mu_1 = \mu_2 = \ldots = \mu_k$ is computed by the F-test:

$$F_o = MSA/MSE \sim \mathcal{F}_{k-1,N-k}$$

where MSA = SSA/$(k - 1)$ and MSE = SSE/$(N - k)$ The $p-$value for the F-test therefore provides us with the "overall" conclusion for the ANOVA test.

ANOVA Table Statistical software such as JMP will present the ANOVA results as:

Source	df	SS	MS	F-Value	P-Value
TREATMENT	$k-1$	SSA	MSA	F_o	$P[F > F_o]$
ERROR	$N-k$	SSE	MSE		
C-TOTAL	$N-1$	SST			

where a significant result (reject H_o) $\Rightarrow$ $p-$value $= P[F > F_o] < 0.05$

One-way ANOVA Assumptions

1. The k groups are independent random samples

[1]The cross product term $\sum_{ij}(\overline{y}_i - \overline{y})(y_{ij} - \overline{y}_i) = \sum_i(\overline{y}_i - \overline{y}) \sum_j(y_{ij} - \overline{y}_i) = \sum_i(\overline{y}_i - \overline{y})[n_i\overline{y}_i - n_i\overline{y}_i] = 0$

2. The group samples are each from a Normal distributed population
3. The populations have equal variation: $\sigma_1^2 = \sigma_2^2 = \ldots = \sigma_k^2$

For the ANOVA results to be considered valid the assumptions above need to be satisfied. Assumption 1 is equivalent to selecting *names-from-a-hat* where, in the Corn example, the first five PLOTs selected from $N = 20$ are assigned to the Control NITROGEN level, the second five PLOTs assigned to the 5% NITROGEN level, and so on. This is called a *Completely Randomized Design* (CRD).

Assumption 2 would be assessed from the *Residuals* using a Q-Q plot, Shapiro-Wilk test, and heuristically from a histogram. Since N may be relatively small, the main concern is whether there is obvious skewness or outliers in the residual diagnostics. Transformations to symmetry may be needed if the residual distribution is clearly nonnormal.

Equal variances is the more critical assumption. If the variances differ greatly from one group to another, the computed significance level ($p-$value) of the F-test may be larger or smaller than the actual significance level realized. That is, we reject H_o if the $p-$value is less than 0.05, but the actual $p-$value is 0.12 when the largest to smallest group variance ratio is 7:1 with equal replications (number of PLOTs per group). We can formally test Assumption 3 which is generally satisfied when the group variance ratio is 3:1 or less.

In practice, minor departures from the ANOVA model assumptions will still provide valid inferences. Gross departures however, particularly unequal group variances with unequal replications, can seriously affect statistical inferences and our conclusions. A nonparametric alternative called the Kruskal-Wallis test can be used in these situations.

JMP Output

For the Corn example, we can use the **Fit Y by X** platform after reformatting the data into two columns: **Nitrogen** (change the Modeling Type to Nominal or Ordinal) and **Yield** In the **Fit Y by X** dialog window, enter **Nitrogen** as the X, Factor and **Yield** as the Y, Response. Click **OK**. Figure 10.1 shows the Report window with the popup menu options.

The **Display Options** was used to add Boxplots and then **Means and Std Dev** was selected as shown in Figure 10.2 There are no outliers, the ranges of the Boxplots look similar, and the distributions for each group appear symmetric. From this exploratory analysis, it is reasonable to perform an ANOVA by selecting **Means / Anova** from the popup menu. The ANOVA Table output is shown in Figure 10.3 and indicates the result is strongly significant: $p-$value =**Prob** $>$ **F** < 0.0001

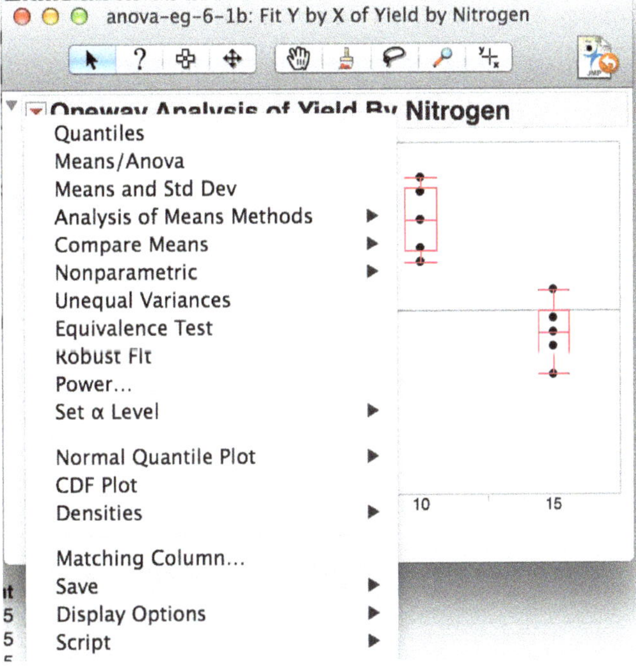

FIGURE 10.1. JMP Options for One-way ANOVA

10.2 ANOVA Assumptions

Before continuing with the post hoc analysis of the group means, we need to check the ANOVA model assumptions which can be tersely stated as $\epsilon_{ij} \overset{\text{i.i.d.}}{\sim} N(0, \sigma^2)$ That is

1. Independent random samples
2. Normal distribution
3. Equal variances

Independent random samples would be satisfied based on the "simple" description given for the Corn study. This assumes a complete randomization of `NITROGEN` levels to `PLOT`s and homogeneous conditions (soil type, irrigation, harvest time, etc.). In practice, other "factors" may need to be included in the analysis to control for known sources of variation.

Normality and Equal Variances can be formally tested. For Normality, select **Save** → **Save Residuals** and run the Shapiro-Wilk test on the saved *residuals* using the **Distribution** platform: **Continuous Fit** → **Normal** and then select **Goodness of Fit** from the popup menu in the **Fitted Normal** report. A nonsignificant $p-$value indicates that the Normality assumption can be accepted. **Normal Quantile Plot** provides a $Q-Q$ plot.

The equal variances assumption is more critical and there are several[2] formal tests available. Hartley's test based on $F_{\max} = s^2_{\max}/s^2_{\min}$ (where $s^2_{\max}, s^2_{\min}$ are the maximum and

[2] Conover et al. (1981)compared 56 different tests for homogeneity of variances!

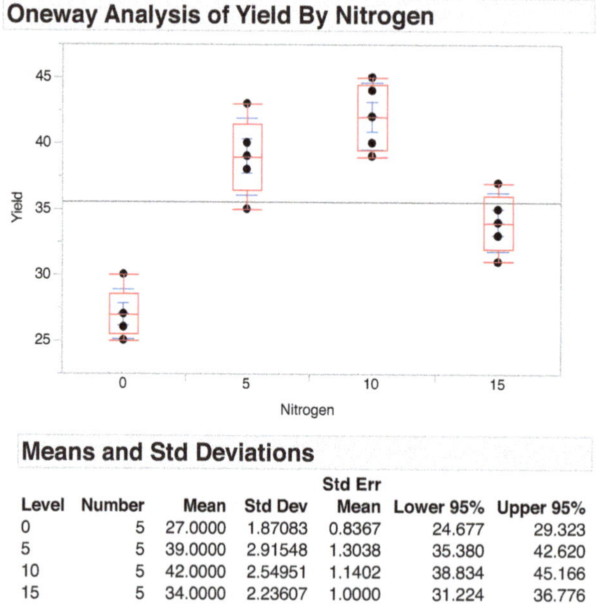

Level	Number	Mean	Std Dev	Std Err Mean	Lower 95%	Upper 95%
0	5	27.0000	1.87083	0.8367	24.677	29.323
5	5	39.0000	2.91548	1.3038	35.380	42.620
10	5	42.0000	2.54951	1.1402	38.834	45.166
15	5	34.0000	2.23607	1.0000	31.224	36.776

FIGURE 10.2. Corn Example: Boxplots / Means and Std Dev

minimum within group sample variances) would seem like an intuitively obvious choice, but this test is sensitive to both non-normality and unequal replications across groups. **Levene's** test does not require Normality and is based based on a one-way ANOVA of $z_{ij} = |\, y_{ij} - \tilde{y}_i \,|$ where $\tilde{y}_i$ is the *median* of the i'th group

In the ANOVA report window, select **Unequal Variances** and check the $p-$value for the Levene test. A nonsignificant value indicates the equal variances assumption can be accepted. Results are shown in Figures 10.4 and 10.5. Both Normality and Equal variances are satisfied for the Corn data example.

When the ANOVA assumptions are NOT satisfied

Independent Samples

The statistical theory underlying the validity of our conclusions (inference) from confidence intervals and hypothesis tests has been based on the condition the sample was randomly selected and that the observations are independent of one another. While random selection does not guarantee any particular sample is necessarily representative of the target population, the ensemble of many samples can be shown to be representative. This, of course, is the risk we must take by only measuring part of the population. However, the observations from a random sample may not be independent if they are correlated with other factors not accounted for in the design of the study. For example, suppose in the Corn study the soil type differs systematically from west to east and all the plots "randomly" assigned to a particular Nitrogen level were on the eastern side. Hence, the Yield cannot be attributed only to effect of this Nitrogen level since it is potentially *confounded* with Soil type.

Oneway Analysis of Yield By Nitrogen

Oneway Anova

Summary of Fit

Rsquare	0.872801
Adj Rsquare	0.848951
Root Mean Square Error	2.42384
Mean of Response	35.5
Observations (or Sum Wgts)	20

Analysis of Variance

Source	DF	Sum of Squares	Mean Square	F Ratio	Prob > F
Nitrogen	3	645.00000	215.000	36.5957	<.0001*
Error	16	94.00000	5.875		
C. Total	19	739.00000			

FIGURE 10.3. Corn Example: ANOVA Table

Key Point Lack of independent observations may result in conclusions that are misleading.

It follows that even if we are only interested in one particular `TREATMENT` factor, the one-way ANOVA model with complete randomization would not be appropriate if the experimental units are heterogeneous due to other extraneous factors (Soil Type) and physical relationships that exist between the units (Plot location). Careful *planning* and implementation of an experiment are therefore critical.

Design of Experiments is a specialized area of statistics that addresses a wide range of analysis methods for more complex experiments. This includes having several factors as well as longitudinal and repeated measures studies where the same subject is measured multiple times (so the responses are correlated). Although this topic is beyond the scope of our text, the bottom line is that not much may be able to be salvaged from a **"botched"** experiment and the results may simply have to be discarded!

Key Point Use the **60-10-30** effort rule: 60% plan ; 10% analysis ; 30% report writing

Normality

The general consensus is that Equal Variances take precedence over Normality. That is, if Levene's test suggest equal group variances is satisfied (nonsignificant $p-$value), but the Shapiro-Wilk test is significant (non-normality of the residuals), the one-way ANOVA results should still be reliable.

Non-normality may be due to the small sample sizes employed in designed experiments, outliers, or lack of equal group variances. A transformation of the response Y may help stabilize the variances and induce Normality. Alternatively, the nonparametric Kruskal-Wallis test can be employed.

Kruskal-Wallis Test Rank the responses y_{ij} then run a one-way ANOVA on the Ranks.

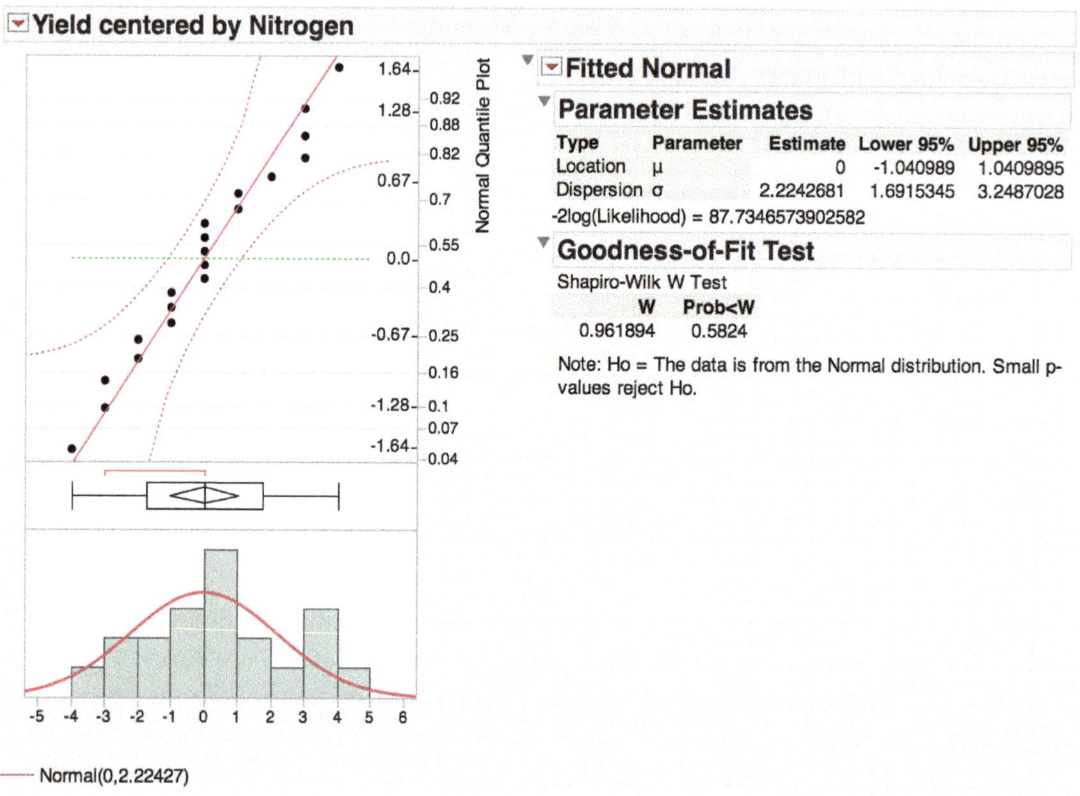

FIGURE 10.4. Corn Example: Normality

Unequal Variances

As indicated above, a power transformation of the response Y can sometimes be used to stabilize the variance across groups. Three common situations are where Y represents:

- **Counts** Use $Y_T = \sqrt{Y}$ or $\sqrt{Y + 3/8}$
- **Populations** Use $Y_T = \log(Y)$ or $\log(Y + 1)$ if $Y = 0$ values exist
- **Proportions** Use $Y_T = \arcsin(\sqrt{Y}) = \sin^{-1}\left[\sqrt{Y}\right]$

Excluding an "outlier" may also fix an unequal variances problem. However, simply excluding the data you don't "like" is not sufficient justification! Run the Kruskal-Wallis test.

10.3 Multiple Comparisons

The results of your study were significant and passed the one-way ANOVA model assumptions. Now we want to do a Post-hoc analysis to determine which levels of `TREATMENT` factor differ significantly. Can we now look at all *pairwise* comparisons? Unless $k = 2$, the answer is no because the Type I error for multiple comparisons is inflated. To see why, suppose each

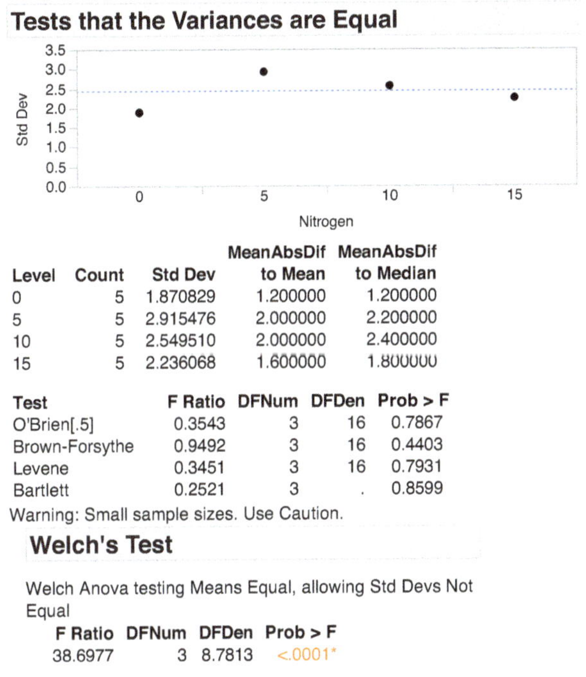

Level	Count	Std Dev	MeanAbsDif to Mean	MeanAbsDif to Median
0	5	1.870829	1.200000	1.200000
5	5	2.915476	2.000000	2.200000
10	5	2.549510	2.000000	2.400000
15	5	2.236068	1.600000	1.800000

Test	F Ratio	DFNum	DFDen	Prob > F
O'Brien[.5]	0.3543	3	16	0.7867
Brown-Forsythe	0.9492	3	16	0.4403
Levene	0.3451	3	16	0.7931
Bartlett	0.2521	3	.	0.8599

Warning: Small sample sizes. Use Caution.

Welch's Test

Welch Anova testing Means Equal, allowing Std Devs Not Equal

F Ratio	DFNum	DFDen	Prob > F
38.6977	3	8.7813	<.0001*

FIGURE 10.5. Corn Example: Equal Variances

pairwise $t-$test is conducted at the α (= 0.05 say) significance level and that the TREATMENT factor has $k > 2$ groups. Then there are a total of $k(k-1)/2$ *pairwise* $t-$tests that could be conducted. For example, in the introduction to this chapter we showed that when $k = 4$, there were 6 ($= 4 \times 3/2$) possible pairwise $t-$tests. And the number of pairwise comparisons clearly grows rapidly as k increases: $k = 10 \Rightarrow 45$ "pairs."

While the probability of *falsely* rejecting any **specific** "pair" is still α (Type I error), the overall experiment-wise Type I error, α_E say, of *falsely* rejecting *any* pair among $m \geq 1$ selected multiple pairwise comparisons could be as high as

$$P[\text{at least one } H_o \text{ falsely rejected }] = 1 - P[\text{none of the } H_o \text{ are rejected }] = 1 - (1-\alpha)^m$$

which, for $m = 10$, is 0.401 when $\alpha = 0.05$ Thus, if 10 pairwise comparisons are made at the individual 5% significance level, then the experiment-wise significance level α_E could actually be as high as 40% — meaning, we have up to a 40% chance of falsely concluding there is at least one significant difference between the TREATMENT factor levels when, in fact, *none* of the $m = 10$ selected pairwise $t-$tests are significant !

Now we have the situation where the overall ANOVA analysis showed there were significant differences between some of the TREATMENT factor levels, but the result doesn't tell us which specific levels differ significantly — and pairwise $t-$tests can inflate α_E to the extent that our conclusions could be very misleading. In the Corn example for example, there is evidence that suggests adding NITROGEN helps increase YIELD, but from a statistical perspective we would like to know which level(s) of NITROGEN "significantly" increase YIELD (on average) since some plots with 5% did better than those with 10%.

To determine the significant differences between the NITROGEN levels we can use the option **Compare Means → Tukey's HSD** The result is shown in Figure 10.6. This shows that the 5% and 10% are *not* significantly different (same Letter), but both these NITROGEN levels are significantly from 0% and 15% which are also significantly different from each other.

Comparisons for all pairs using Tukey-Kramer HSD

Connecting Letters Report

Level				Mean
10	A			42.000000
5	A			39.000000
15		B		34.000000
0			C	27.000000

Levels not connected by same letter are significantly different.

FIGURE 10.6. Corn Example: Multiple Comparisons

The Tukey multiple comparison procedure controls the experiment-wise Type I error α_E by using the statistic $(\overline{y}_{\max} - \overline{y}_{\min})$ where $\overline{y}_{\max}$ and $\overline{y}_{\min}$ denote the largest and smallest group means (NITROGEN level means in the example). Hence any other "pairwise" comparison lies within this range. Thus, the group means $\overline{y}_i$ and $\overline{y}_j$ are declared significantly different if

$$|\overline{y}_i - \overline{y}_j| > q_{\alpha_E,k,\nu}\sqrt{\text{MSE}/n}$$

where k = number of groups (NITROGEN levels) and ν = error degrees-of-freedom (MSE), controlled for $\alpha_E(= 0.05)$ where $q_{\alpha_E,k,\nu}$ is tabulated in Appendix B (Studentized Range).

10.4 Linear Contrasts

In the previous sections we discussed the following hypotheses

$$H_o : \mu_1 = \mu_2 = \mu_3 = \mu_4 \quad \text{(ANOVA)} \quad \text{and} \quad H_o : \mu_j = \mu_k \quad \text{(Pairwise Comparisons)}$$

and the Tukey HSD multiple comparison method. If the the ANOVA test is significant the researcher may also be interested in other objectives. For example, In the Corn study "does adding any amount of NITROGEN increase YIELD" (as compared to the 0% level).

Translating this *directional* objective to the NULL case — adding any amount of NITROGEN makes no significant difference to YIELD — we can express this in word form as:

$$average(5\%,\ 10\%,\ 15\%)\ -\ average(0\%) = 0$$

While this might seem reasonable, it actually represents a "new" experiment wherein the non-zero levels of NITROGEN (5%, 10%, 15%) would need to have been randomly assigned to the other 15 PLOTs. That is, this "new" experiment "could" allow 10 PLOTs to be randomly assigned 5% and the other five (5) PLOTs only assigned to 10%, or 15%, or split between these levels. Clearly, this was not how the original experiment was implemented as each level of NITROGEN was specifically assigned 5 PLOTs.

Using the Corn study, we can formulate the researcher's question as a "Linear Contrast" estimated by: $\frac{1}{3}(\overline{y}_{5\%} + \overline{y}_{10\%} + \overline{y}_{15\%}) - \overline{y}_{0\%}$ Under the null hypothesis this is

$$C1 : \frac{1}{3}(\mu_2 + \mu_3 + \mu_4) - \mu_1 = 0$$

which can be coded as $C1 = (-1, \frac{1}{3}, \frac{1}{3}, \frac{1}{3})$ or equivalently as $C1 = (-3, 1, 1, 1)$ (since there are equal replications per group). Similarly, another contrast of interest might be whether $(0\% + 15\%)$ differ from $(5\% + 10\%)$ Again, we would use the null hypothesis

$$C2 : \frac{1}{2}(\mu_1 + \mu_4) - \frac{1}{2}(\mu_2 + \mu_3) = 0$$

which can be coded as $C2 = (\frac{1}{2}, \frac{-1}{2}, \frac{-1}{2}, \frac{1}{2})$ or equivalently as $C2 = (1, -1, -1, 1)$

Key Point: The sum of the Contrast Coefficients must equal 0

That is, any linear contrast $C : a_1\mu_1 + a_2\mu_2 + \cdots + a_k\mu_k$ must satisfy the requirement that $\sum_{i=1}^{k} a_i = 0$ so we are testing the null hypothesis $H_o : C = 0$ Note that C is not unique since we could have expressed $C1$ as $C1 : \mu_1 - \frac{1}{3}(\mu_2 + \mu_3 + \mu_4)$

Example 10.1 Does Exercise and/or Counseling help 50 overweight males weight-loss regime in addition to administering a weight-loss drug? See study details below

Ten male subjects were randomly assigned to each of the following Treatment groups:

S Standard therapy (Control: No Drug)

A1 Drug with Exercise and Counseling

A2 Drug with Exercise only

A3 Drug with Counseling only

A4 Drug only

After removing a high outlier in the S group (weight loss = 12.2), the overall ANOVA was significant ($p-$value $< .0001$) and Levene's test was not significant ($p-$value $= 0.5421$). However, the Tukey HSD report shown in Figure 10.8 was somewhat ambiguous! This seemed to suggest Exercise and/or Counseling only had a minor effect versus just administering the Drug. (The new Treatments were all significantly better than the S group.) For the researcher, Linear Contrasts can help answer questions such as:

1. Did the new Treatments ($A1, A2, A3, A4$) on average, do better than S ?

   $C1 : \mu_{A1} + \mu_{A2} + \mu_{A3} + \mu_{A4} - 4\mu_S = 0$ $\quad C1 = (1, 1, 1, 1, -4)$

2. Did Counseling help versus non-Counseling? (Ignoring S)

   $C2 : \mu_{A1} + \mu_{A3} - (\mu_{A2} + \mu_{A4}) = 0$ $\quad C2 = (1, -1, 1, -1, 0)$

Comparisons for all pairs using Tukey-Kramer HSD

Connecting Letters Report

Level					Mean
A4	A				12.170000
A1	A	B			12.050000
A2		B	C		11.020000
A3			C		10.270000
S				D	8.944444

Levels not connected by same letter are significantly different.

FIGURE 10.7. Weight Loss Example: Tukey HSD

3. Did Exercise help versus non-Exercise? (Ignoring S)

$$C2 : \mu_{A1} + \mu_{A2} - (\mu_{A3} + \mu_{A4}) = 0 \qquad C3 = (1, 1, -1, -1, 0)$$

To test these contrasts, the **Analyze –> Fit Model** platform can be used. To do this, first conduct the ANOVA analysis: Enter **Agent** as the X, Factor and **wgt loss** as the Y, Response. Click **OK**. From the report window, select the popup menu by **Agent** and choose LS Means Contrast. Now click on the "+" icon for each of $A1$ to $A4$ and the "-" icon for S. JMP automatically rescales the $C1$ contrast to $(0.25, 0.25, 0.25, -1)$ Now click Done. Your result will be:

Contrast

Test Detail

A1	0.25
A2	0.25
A3	0.25
A4	0.25
S	-1
Estimate	2.4331
Std Error	0.3298
t Ratio	7.3769
Prob>|t|	3.2e-9
SS	43.492

SS	NumDF	DenDF	F Ratio	Prob > F
43.49	1	44	54.4189	<.0001*

FIGURE 10.8. Weight Loss Example: Contrast to S

The $C2$ and $C3$ can be done from the same report window in similar fashion. Both are not significant which suggests Exercise and/or Counseling do not make a significant impact beyond administering the weight-loss Drug.

Key Point: Designing an "effective" study is not easy.

Chapter 10 Exercises

10.1 The response time in milliseconds was determined for three different types of circuits used in an automatic shutoff mechanism.

a) Create a JMP dataset from Table 10.2
b) Perform the ANOVA test
c) State your conclusion
d) What is unusual about the last response times for each circuit? Explain.

TABLE 10.2. Circuit Shutoff Times (ms)

Circuit Type	Response Time				
1	9	12	10	8	15
2	20	21	23	17	30
3	6	5	8	7	16

10.2 PANES Y = Heatloss from five types of window panes A, B, C, D, E at five external temperature levels 0, 20, 40, 60, 80 in °F

a) Levenes test is significant (p–value = 0.0091) for the Y vs Pane ANOVA analysis. Explain what this implies about the Y versus Pane ANOVA analysis.
b) Explain why the overall ANOVA analysis for the Y vs ExtTemp is significant
c) Interpret the Tukey HSD result.

10.3 DOGS Two different Diets (D1 and D2) will be used for investigating the health of 16 dogs. All dogs are in good health and at the end of three months the cardiovascular health of the dogs is to be measured. The dataset provides information on the Gender and Age of the 16 dogs. The aim is to "design" the planned investigation by assigning each dog to one of the Diets so that Gender AND Age are equally distributed (balanced) across both Diets

a) If Age is ignored, argue that the Diet allocation could be badly biased.
b) Do a double-sort of the dogs dataset.Based on the sorted dataset, a *balanced* allocation of Gender AND Age cannot be obtained for each Diet. Explain why.
c) By grouping (categorizing) Age show your best design for this study

Appendix A

Statistical Software Notes

A.1 Software Packages

There is a wide variety of software packages that can be used for statistical analysis. Our list is by no means comprehensive, nor does it represent an endorsement of any of these products. However, it is worthwhile comparing some general features of these packages. Table A.1 provides a subjective classification of some standard statistical software packages.

For Mac OS users, a native Windows partition (via Bootcamp), or a Virtual Windows application, is required to run some of these software packages (e.g., SAS). Currently, Excel, JMP, and R provide versions for both Mac OS and Windows. For this reason, JMP is used throughout this book.

TABLE A.1. Statistical Software

Level	Package	Features
Basic	Calculators	The TI (Texas Instruments) series of calculators are used extensively by Math departments and can provide basic statistical analysis with graphical displays.
	Excel	Provides the basic statistical analyses: EDA, Regression, t-tests, etc. If this is all you need, use it.
Intermediate	JMP	Menu driven. Pro version is comprehensive.
	Minitab	A favorite of Business schools. Mostly menu driven.
	Textbook	Some texts include their own Statistical software package (for an extra fee) such as StatCrunch, CrunchIt!
	Website	SAS OnDemand and R Studio provide browser access Useful for Mac OS users (no Windows VM required).
	Python	Similar to R. Emphasis on Applied Math. Code based
Comprehensive	SAS	Expensive. Steep learning curve. Code based.
	SPSS	Menu driven and code based. Used extensively in the Social Sciences. Data format can be a little quirky.
	R	Not user-friendly. Excellent graphics. Requires extensive coding and knowledge by the user.
Specialized	GLIM	Generalized LInear Models. User-knowledge assumed.
	Stata	Exact power speciality. Comprehensive suite.

A.2 JMP

JMP (pronounced as "jump") is a menu-driven statistical software package that is available for the Mac OS and Windows platforms. The current version and system requirements can be found at: `http://www.jmp.com/support/system_requirements_jmp.shtml` Additional resources can be found at: `http://www.jmp.com/software/jmp/#Key-Features`

A.2.1 JMP Data Table

There are several components involved when you use JMP so we first define some basic terminology for consistency. See Figure A.1 for reference.

Data Table This is an existing dataset which is in JMP format. That is, you can just double-click on its icon to start JMP and open the dataset within the JMP application.

Data Table Submenus In the top left corner of the Data Table there are three triangles around a diagonal south-east sloping line.

- Blue (grey on a Mac) opens/closes the Side Panel which provides information about the Column variables and Row attributes.

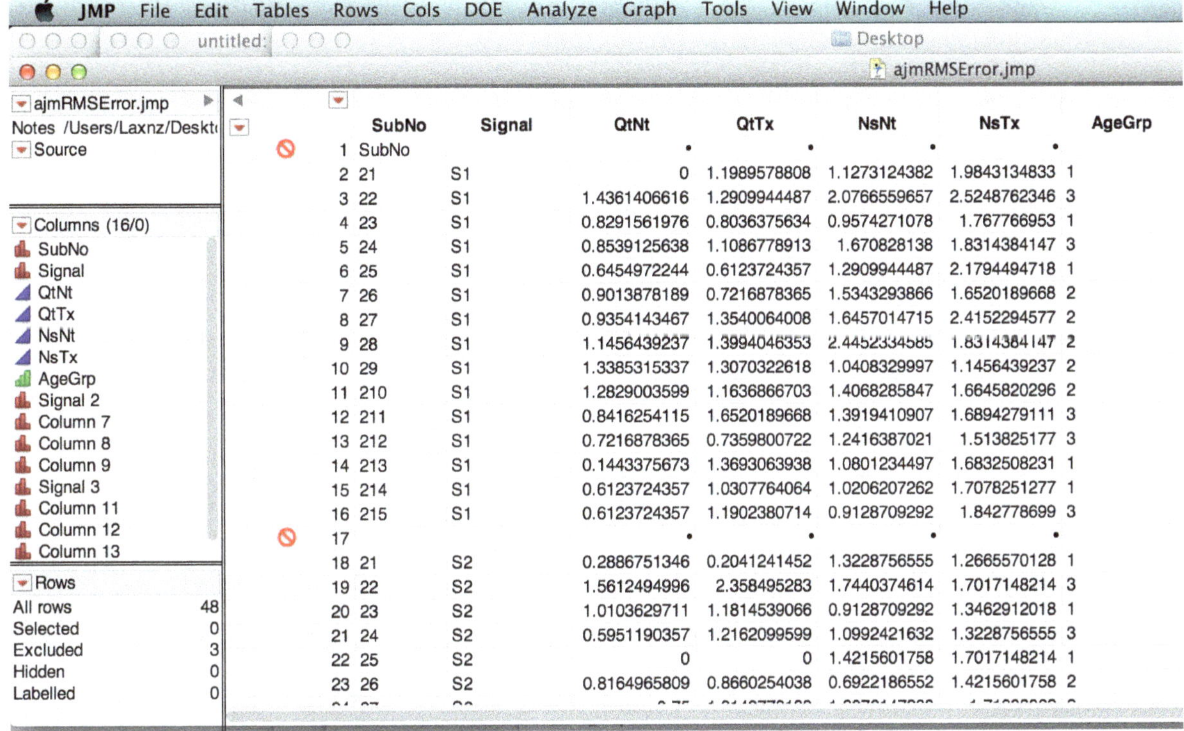

FIGURE A.1. JMP Screen Shot (Mac version)

- Red (top right) is a pulldown submenu which duplicates the Col options in the Top menu (see below)
- Red (bottom left) is a pulldown submenu which duplicates the Row options in the Top menu (see below)

  To unhighlight column(s), row(s), or cell(s) in a Data Table, click above and/or below the sloping line as needed.

Top menu This refers to the pulldown menus that appear at the top of the JMP application window: **File Edit Tables ...** (Mac JMP version shown in Figure A.1)

To select an option from one of these menus, click-and-hold, slide the mouse down to the desired option, then release

Popup Window Selecting some options from a pulldown menu (e.g., Analyze → Distribution) will result in a popup window that allows you to perform certain tasks, such as selecting the variables to be analyzed, that will be carried out once you click OK. The popup window will then disappear.

Output Window The output generated by JMP appears in a separate window that you can print. However, do NOT use JMP to save the output to a file. See Saving JMP Output below

Note: *Once you close this output window it is gone. You would then need to repeat all the steps you performed to recreate this output.*

Reports The Output window will contain one more "Reports" which may be graphs, numerical summaries, or the results of a statistical analysis.

Clicking on the Blue triangle beside a particular report will close that report (but not the Output window) and will not appear on the printout. To open a particular report that is closed, click on the Blue triangle. Some other reports have Red triangles which provide popup Submenu(s) that enable additional analyses to be performed and the associated report appended to the current Output window.

Submenu A popup menu that provides options to be selected in the same way as the Top menu. The Submenu options depend on the type of window in which they appear.

A.2.2 Saving JMP Output

Printing a JMP Output window is no problem: **File** → **Print**. Saving JMP Output to a file is a major deficiency of this software package. Using **File** → **Save** (or **Save As ...**) within JMP creates a `*.jrp` file which is a "script" file — a plain text file containing the JMP menu option selections you used to create the output — *not* the output image itself. That is, the recipient of your emailed `*.jrp` file must have (1) an active version of JMP and (2) your dataset to run the `*.jrp` script. Very inefficient. Two work arounds are:

Print to File Under OSX (Mac) you can save the Output to a pdf file directly by selecting the PDF option in the Print window. This can be done even if you have no printer attached to your Mac. Similar "save-to-file" options are available in the Print menu on PC Windows.

Saving JMP Output to Word Open up a Word document where you want the JMP Output to go. In JMP, with the Output window active (if not, click on its title bar), then select: Edit → Copy Now go back to the Word document, click where you want the JMP output to go, and select: Edit →Paste The JMP Output should appear after a brief pause. Don't forget to save the Word document.

A.2.3 Data Types

JMP depends critically on the data type of variables to produce the correct analysis. For example, if you type in 1 and 2 as coded entries for male and female, JMP will think the "Gender" column is numeric. The three (3) classifications for a variable "X" are:

Continuous X is a quantitative variable so both $x_i < x_j$ and $x_i - x_j$ make sense. That is, order and differences in magnitude are valid comparisons between observations.

Denoted by a Blue triangle beside the variable name in the Side Panel e.g., `QtNt`

Ordinal X has specific levels that have an inherent ranking e.g., Low, Medium, High but the magnitude of the difference Low - High is not meaningful.

Denoted by a Green bar chart e.g., `AgeGrp`

Nominal X is simply a "label" of the levels of X (e.g., Male / Female).

Denoted by a Red bar chart e.g., `Signal`

To coerce a specific data type for a variable, click on the column header of X, then select **Col → Col Info ...** from the top menu. In the popup window the options are:

Data Type Numeric or Character (ignore Row State)

Modeling Type Continuous, Ordinal, or Nominal

Click **Apply** to check the result without closing the popup window or click **OK** to make the changes and close the popup window. As can be seen in Figure A.1, Character variables have left-flush entries and Numeric variables have right-flush entries. Hence the Data Type of `AgeGrp` is currently treated as a "character" in Figure A.1. If we change its Data Type to "numeric" the entries will become right-flush, but its Modeling Type would remain as Ordinal (unless we change this as well).

A.2.4 Using JMP: Basics

For univariate and bivariate analyses, our two main "platforms" are **Distribution** and **Fit Y by X**. Both platforms are accessed via **Analyze** in the Top menu. For convenience, let A, B represent two categorical variables (eg. gender and marital status), and X, Y represent two quantitative variables (eg. GPA and Income).

Distribution Provides bar graphs for categorical variables, and histograms for quantitative variables. To use the distribution platform do:

Select **Analyze → Distribution** In the popup window that appears, click on the variable name in the left panel, then on **Y, Variables** to add this variable to the right panel. You can add as many variables as you like. When the variables you want to analyze have been added, click **OK**. A report window will appear with bar graphs and/or histograms depending on what variables you added.

Beside the title "Distribution" is a red triangle. This popup menu enables you switch between horizontal and vertical display formats by selecting **Stack** — reselecting **Stack** changes the display format back.

Beside each "variable name" is another red triangle which provides more options:

Display Options Change (this variable only) to horizontal layout (or vice-versa by reselecting). Customize Summary Statistics — add desired statistics.

Histogram Options Add a count or probability axis to the graph.

Stem and Leaf Add a "Stemplot" display to the output window.

Test Mean Perform a one-sample $z-$test (need to specify $\sigma =$ `Sigma`), or $t-$test (do NOT specify σ).

Fit Y by X Select **Analyze** → **Fit Y by X** The four bivariate options are:

Y vs A Parallel dot plot: two-sample $t-$test if A has two levels.
Oneway ANOVA if A has more than two levels. See Using JMP: Advanced

B vs A Contingency table analysis.

Y vs X Scatter plot.

B vs X Logistic regression.

Two-sample $t-$test

Enter the *two-level* A variable (listed in the left panel) as the **X, Factor** then enter the continuous Y variable as the **Y, Response**. Click **OK**

The output window will show a parallel "dot" plot of Y for each level of A. Use the red triangle popup menu **Display Options** → **Box Plots** to add boxplots to this display. Select via the red triangle **$t-$test** to get the two-sample t-test.

Contingency Tables

Enter the A variable as the **X, Factor** then enter the B variable as the **Y, Response**. Click **OK** [Note: you can switch A and B here]

The report window will show a"mosaic" plot and a contingency table of the frequencies (counts) of A versus B. There will also be a test report listing the Pearson Chi square test.

Regression

Enter the X variable as the **X, Factor** then enter the Y variable as the **Y, Response**. Click **OK**

The output window will show a scatter plot of Y vs X. Some options are:

Fit Line Add a SLR line to the display. To get separate SLRs fitted to the same scatter plot select **Group By ...**

Density Ellipse Get the correlation between X and Y To show the correlation report click on the blue triangle beside the title "Correlation."

Logistic Regression

Enter the X variable as the **X, Factor** then enter the B variable as the **Y, Response**. Click **OK**

Note: If you expected to do a $t-$test (or oneway ANOVA) then you entered the X and B variables in the wrong places.

A.2.5 Top Menu Options

File → Open Enables you to open an existing JMP file, or create one from another file such as an Excel spreadsheet.

Note You may need to convert *.xlsx files to *.xls format for earlier versions of JMP

File → Save / Save As ... "Save" enables you to save changes you make to an existing JMP file. Use "Save As ..." if you want to save the JMP file under a new name.

Note Do NOT use this to save Output window results. See Saving JMP Output above.

File → New Enables you to create a JMP dataset. The default is a JMP dataset with one column called **Column 1** and blank row entries. The following describes how to enter data, add and rename columns, and add rows.

Enter Data Simply type in your data entries using the Return / Enter key (or Down Arrow key) to move down to a new cell.

If you start with a "character" entry JMP will automatically convert **Column 1** to Data Type = Character with Modeling Type = Nominal.

If you start with a "numeric" entry, JMP assumes Data Type = Numeric with Modeling Type = Continuous. If you later enter a "character" value, a popup window appears asking whether you want to change **Column 1** to Data Type = Character. Click on the appropriate response.

Add Columns To add column(s) manually, double-click on the empty space beside the **Column 1** header. This will create a new column named **Column 2**. Repeat as needed. Alternatively, use the Top menu **Cols → Add Multiple Columns ...** wherein you can preset the Column prefix, How many columns to add, Data Type, and where to add the new columns (i.e., new variables)

Rename Columns Click in the header space of the column you want to rename. Double-click to allow you type in your desired name. Alternatively, highlight all the columns you want rename and use **Cols → Column Info...** to rename each selected column.

Add Rows From the Top menu use **Rows → Add Rows ...** to add as many rows as needed. You can also use the Rows menu to delete rows by highlighting the row(s) you want to delete.

Cols Add new columns, delete columns, or move columns (variables)

Rows Add rows, delete rows, exclude rows, add markers, and select subsets of the data.

Tables There are several useful reformatting options available in this JMP menu.

Appendix B
Statistical Tables

We note that some scientific calculators can compute the probabilities associated with the statistical tables listed below. Statistical software packages such as R have a wide range of distributions. For example `pnorm(2)` $= 0.9772499 = P[Z < 2]$

Applets such as https://homepage.divms.uiowa.edu/~mbognar/applets/ are also available for a variety of distributions Bognar (2021)

For completeness, we provide the following statistical tables

Table A.1	Standard Normal Distribution ($z < 0$)
Table A.2	Standard Normal Distribution ($z \geq 0$)
Table B	t−Distribution
Table C	Chi-square Distribution
Table D	$\mathcal{F}$−Distribution $\alpha = 0.05$
Table E	Studentized Range

Table A.1 Standard Normal Distribution. Table entry is $P[\,Z < z\,]$

z	0.00	0.01	0.02	0.03	0.04	0.05	0.06	0.07	0.08	0.09
-3.4	0.0003	0.0003	0.0003	0.0003	0.0003	0.0003	0.0003	0.0003	0.0003	0.0002
-3.3	0.0005	0.0005	0.0005	0.0004	0.0004	0.0004	0.0004	0.0004	0.0004	0.0003
-3.2	0.0007	0.0007	0.0006	0.0006	0.0006	0.0006	0.0006	0.0005	0.0005	0.0005
-3.1	0.0010	0.0009	0.0009	0.0009	0.0008	0.0008	0.0008	0.0008	0.0007	0.0007
-3.0	0.0013	0.0013	0.0013	0.0012	0.0012	0.0011	0.0011	0.0011	0.0010	0.0010
-2.9	0.0019	0.0018	0.0018	0.0017	0.0016	0.0016	0.0015	0.0015	0.0014	0.0014
-2.8	0.0026	0.0025	0.0024	0.0023	0.0023	0.0022	0.0021	0.0021	0.0020	0.0019
-2.7	0.0035	0.0034	0.0033	0.0032	0.0031	0.0030	0.0029	0.0028	0.0027	0.0026
-2.6	0.0047	0.0045	0.0044	0.0043	0.0041	0.0040	0.0039	0.0038	0.0037	0.0036
-2.5	0.0062	0.0060	0.0059	0.0057	0.0055	0.0054	0.0052	0.0051	0.0049	0.0048
-2.4	0.0082	0.0080	0.0078	0.0075	0.0073	0.0071	0.0069	0.0068	0.0066	0.0064
-2.3	0.0107	0.0104	0.0102	0.0099	0.0096	0.0094	0.0091	0.0089	0.0087	0.0084
-2.2	0.0139	0.0136	0.0132	0.0129	0.0125	0.0122	0.0119	0.0116	0.0113	0.0110
-2.1	0.0179	0.0174	0.0170	0.0166	0.0162	0.0158	0.0154	0.0150	0.0146	0.0143
-2.0	0.0228	0.0222	0.0217	0.0212	0.0207	0.0202	0.0197	0.0192	0.0188	0.0183
-1.9	0.0287	0.0281	0.0274	0.0268	0.0262	0.0256	0.0250	0.0244	0.0239	0.0233
-1.8	0.0359	0.0351	0.0344	0.0336	0.0329	0.0322	0.0314	0.0307	0.0301	0.0294
-1.7	0.0446	0.0436	0.0427	0.0418	0.0409	0.0401	0.0392	0.0384	0.0375	0.0367
-1.6	0.0548	0.0537	0.0526	0.0516	0.0505	0.0495	0.0485	0.0475	0.0465	0.0455
-1.5	0.0668	0.0655	0.0643	0.0630	0.0618	0.0606	0.0594	0.0582	0.0571	0.0559
-1.4	0.0808	0.0793	0.0778	0.0764	0.0749	0.0735	0.0721	0.0708	0.0694	0.0681
-1.3	0.0968	0.0951	0.0934	0.0918	0.0901	0.0885	0.0869	0.0853	0.0838	0.0823
-1.2	0.1151	0.1131	0.1112	0.1093	0.1075	0.1056	0.1038	0.1020	0.1003	0.0985
-1.1	0.1357	0.1335	0.1314	0.1292	0.1271	0.1251	0.1230	0.1210	0.1190	0.1170
-1.0	0.1587	0.1562	0.1539	0.1515	0.1492	0.1469	0.1446	0.1423	0.1401	0.1379
-0.9	0.1841	0.1814	0.1788	0.1762	0.1736	0.1711	0.1685	0.1660	0.1635	0.1611
-0.8	0.2119	0.2090	0.2061	0.2033	0.2005	0.1977	0.1949	0.1922	0.1894	0.1867
-0.7	0.2420	0.2389	0.2358	0.2327	0.2296	0.2266	0.2236	0.2206	0.2177	0.2148
-0.6	0.2743	0.2709	0.2676	0.2643	0.2611	0.2578	0.2546	0.2514	0.2483	0.2451
-0.5	0.3085	0.3050	0.3015	0.2981	0.2946	0.2912	0.2877	0.2843	0.2810	0.2776
-0.4	0.3446	0.3409	0.3372	0.3336	0.3300	0.3264	0.3228	0.3192	0.3156	0.3121
-0.3	0.3821	0.3783	0.3745	0.3707	0.3669	0.3632	0.3594	0.3557	0.3520	0.3483
-0.2	0.4207	0.4168	0.4129	0.4090	0.4052	0.4013	0.3974	0.3936	0.3897	0.3859
-0.1	0.4602	0.4562	0.4522	0.4483	0.4443	0.4404	0.4364	0.4325	0.4286	0.4247
-0.0	0.5000	0.4960	0.4920	**0.4880**	0.4840	0.4801	0.4761	0.4721	0.4681	0.4641

NOTE: $P[Z < -0.03]$ = (row) -0.0 and (col) $0.03 = 0.4880$ (bold entry)
Similarly $P[Z < -1.96] = 0.0250$

Table A.2 Standard Normal Distribution. Table entry is $P[Z < z]$

z	0.00	0.01	0.02	0.03	0.04	0.05	0.06	0.07	0.08	0.09
0.0	0.5000	0.5040	0.5080	0.5120	0.5160	0.5199	0.5239	0.5279	0.5319	0.5359
0.1	0.5398	0.5438	0.5478	0.5517	0.5557	0.5596	0.5636	0.5675	0.5714	0.5753
0.2	0.5793	0.5832	0.5871	0.5910	0.5948	0.5987	0.6026	0.6064	0.6103	0.6141
0.3	0.6179	0.6217	0.6255	0.6293	0.6331	0.6368	0.6406	0.6443	0.6480	0.6517
0.4	0.6554	0.6591	0.6628	0.6664	0.6700	0.6736	0.6772	0.6808	0.6844	0.6879
0.5	0.6915	0.6950	0.6985	0.7019	0.7054	0.7088	0.7123	0.7157	0.7190	0.7224
0.6	0.7257	0.7291	0.7324	0.7357	0.7389	0.7422	0.7454	0.7486	0.7517	0.7549
0.7	0.7580	0.7611	0.7642	0.7673	0.7704	0.7734	0.7764	0.7794	0.7823	0.7852
0.8	0.7881	0.7910	0.7939	0.7967	0.7995	0.8023	0.8051	0.8078	0.8106	0.8133
0.9	0.8159	0.8186	0.8212	0.8238	0.8264	0.8289	0.8315	0.8340	0.8365	0.8389
1.0	0.8413	0.8438	0.8461	0.8485	0.8508	0.8531	0.8554	0.8577	0.8599	0.8621
1.1	0.8643	0.8665	0.8686	0.8708	0.8729	0.8749	0.8770	0.8790	0.8810	0.8830
1.2	0.8849	0.8869	0.8888	0.8907	0.8925	0.8944	0.8962	0.8980	0.8997	0.9015
1.3	0.9032	0.9049	0.9066	0.9082	0.9099	0.9115	0.9131	0.9147	0.9162	0.9177
1.4	0.9192	0.9207	0.9222	0.9236	0.9251	0.9265	0.9279	0.9292	0.9306	0.9319
1.5	0.9332	0.9345	0.9357	0.9370	0.9382	0.9394	0.9406	0.9418	0.9429	0.9441
1.6	0.9452	0.9463	0.9474	0.9484	0.9495	0.9505	0.9515	0.9525	0.9535	0.9545
1.7	0.9554	0.9564	0.9573	0.9582	0.9591	0.9599	0.9608	0.9616	0.9625	0.9633
1.8	0.9641	0.9649	0.9656	0.9664	0.9671	0.9678	0.9686	0.9693	0.9699	0.9706
1.9	0.9713	0.9719	0.9726	0.9732	0.9738	0.9744	0.9750	0.9756	0.9761	0.9767
2.0	0.9772	0.9778	0.9783	0.9788	0.9793	0.9798	0.9803	0.9808	0.9812	0.9817
2.1	0.9821	0.9826	0.9830	0.9834	0.9838	0.9842	0.9846	0.9850	0.9854	0.9857
2.2	0.9861	0.9864	0.9868	0.9871	0.9875	0.9878	0.9881	0.9884	0.9887	0.9890
2.3	0.9893	0.9896	0.9898	0.9901	0.9904	0.9906	0.9909	0.9911	0.9913	0.9916
2.4	0.9918	0.9920	0.9922	0.9925	0.9927	0.9929	0.9931	0.9932	0.9934	0.9936
2.5	0.9938	0.9940	0.9941	0.9943	0.9945	0.9946	0.9948	0.9949	0.9951	0.9952
2.6	0.9953	0.9955	0.9956	0.9957	0.9959	0.9960	0.9961	0.9962	0.9963	0.9964
2.7	0.9965	0.9966	0.9967	0.9968	0.9969	0.9970	0.9971	0.9972	0.9973	0.9974
2.8	0.9974	0.9975	0.9976	0.9977	0.9977	0.9978	0.9979	0.9979	0.9980	0.9981
2.9	0.9981	0.9982	0.9982	0.9983	0.9984	0.9984	0.9985	0.9985	0.9986	0.9986
3.0	0.9987	0.9987	0.9987	0.9988	0.9988	0.9989	0.9989	0.9989	0.9990	0.9990
3.1	0.9990	0.9991	0.9991	0.9991	0.9992	0.9992	0.9992	0.9992	0.9993	0.9993
3.2	0.9993	0.9993	0.9994	0.9994	0.9994	0.9994	0.9994	0.9995	0.9995	0.9995
3.3	0.9995	0.9995	0.9995	0.9996	0.9996	0.9996	0.9996	0.9996	0.9996	0.9997
3.4	0.9997	0.9997	0.9997	0.9997	0.9997	0.9997	0.9997	0.9997	0.9997	0.9998

Table B t-Distribution
Table entries are t^*_{df} quantiles for Confidence Level C
Corresponding 1-sided and 2-sided $p-$values are shown

	Confidence Level C									
df	50%	60%	80%	90%	95%	96%	98%	99%	99.8%	99.9%
1	1.0000	1.3764	3.0777	6.3138	*12.706*	*15.895*	*31.821*	*63.657*	*318.31*	*636.62*
2	0.8165	1.0607	1.8856	2.9200	4.3027	4.8487	6.9646	9.9248	*22.327*	*31.599*
3	0.7649	0.9785	1.6377	2.3534	3.1824	3.4819	4.5407	5.8409	*10.215*	*12.924*
4	0.7407	0.9410	1.5332	2.1318	2.7764	2.9985	3.7469	4.6041	7.1732	8.6103
5	0.7267	0.9195	1.4759	2.0150	2.5706	2.7565	3.3649	4.0321	5.8934	6.8688
6	0.7176	0.9057	1.4398	1.9432	2.4469	2.6122	3.1427	3.7074	5.2076	5.9588
7	0.7111	0.8960	1.4149	1.8946	2.3646	2.5168	2.9980	3.4995	4.7853	5.4079
8	0.7064	0.8889	1.3968	1.8595	2.3060	2.4490	2.8965	3.3554	4.5008	5.0413
9	0.7027	0.8834	1.3830	1.8331	2.2622	2.3984	2.8214	3.2498	4.2968	4.7809
10	0.6998	0.8791	1.3722	1.8125	2.2281	2.3593	2.7638	3.1693	4.1437	4.5869
11	0.6974	0.8755	1.3634	1.7959	2.2010	2.3281	2.7181	3.1058	4.0247	4.4370
12	0.6955	0.8726	1.3562	1.7823	2.1788	2.3027	2.6810	3.0545	3.9296	4.3178
13	0.6938	0.8702	1.3502	1.7709	2.1604	2.2816	2.6503	3.0123	3.8520	4.2208
14	0.6924	0.8681	1.3450	1.7613	2.1448	2.2638	2.6245	2.9768	3.7874	4.1405
15	0.6912	0.8662	1.3406	1.7531	2.1314	2.2485	2.6025	2.9467	3.7328	4.0728
16	0.6901	0.8647	1.3368	1.7459	2.1199	2.2354	2.5835	2.9208	3.6862	4.0150
17	0.6892	0.8633	1.3334	1.7396	2.1098	2.2238	2.5669	2.8982	3.6458	3.9651
18	0.6884	0.8620	1.3304	1.7341	2.1009	2.2137	2.5524	2.8784	3.6105	3.9216
19	0.6876	0.8610	1.3277	1.7291	2.0930	2.2047	2.5395	2.8609	3.5794	3.8834
20	0.6870	0.8600	1.3253	1.7247	2.0860	2.1967	2.5280	2.8453	3.5518	3.8495
21	0.6864	0.8591	1.3232	1.7207	2.0796	2.1894	2.5176	2.8314	3.5272	3.8193
22	0.6858	0.8583	1.3212	1.7171	2.0739	2.1829	2.5083	2.8188	3.5050	3.7921
23	0.6853	0.8575	1.3195	1.7139	2.0687	2.1770	2.4999	2.8073	3.4850	3.7676
24	0.6848	0.8569	1.3178	1.7109	2.0639	2.1715	2.4922	2.7969	3.4668	3.7454
25	0.6844	0.8562	1.3163	1.7081	2.0595	2.1666	2.4851	2.7874	3.4502	3.7251
30	0.6828	0.8538	1.3104	1.6973	2.0423	2.1470	2.4573	2.7500	3.3852	3.6460
35	0.6816	0.8520	1.3062	1.6896	2.0301	2.1332	2.4377	2.7238	3.3400	3.5911
40	0.6807	0.8507	1.3031	1.6839	2.0211	2.1229	2.4233	2.7045	3.3069	3.5510
50	0.6794	0.8489	1.2987	1.6759	2.0086	2.1087	2.4033	2.6778	3.2614	3.4960
60	0.6786	0.8477	1.2958	1.6706	2.0003	2.0994	2.3901	2.6603	3.2317	3.4602
80	0.6776	0.8461	1.2922	1.6641	1.9901	2.0878	2.3739	2.6387	3.1953	3.4163
100	0.6770	0.8452	1.2901	1.6602	1.9840	2.0809	2.3642	2.6259	3.1737	3.3905
z^*	0.674	0.841	1.282	1.645	1.96	2.054	2.326	2.576	3.091	3.291
1-sided p	0.25	0.20	0.10	0.05	0.025	0.02	0.01	0.005	0.001	0.0005
2-sided p	0.50	0.40	0.20	0.10	0.05	0.04	0.02	0.01	0.002	0.001

Table C Chi-square Distribution.

Table entry is $w^* = \chi_k^2(\alpha)$ where $P[\,W > w^*\,] = \alpha$ (Tail area to the right)

	α = Tail area to the right									
k	0.995	0.99	0.975	0.95	0.9	0.1	0.05	0.025	0.01	0.005
1	3.927(-5)	1.571(-4)	9.821(-4)	0.004	0.016	2.706	3.842	5.024	6.635	7.879
2	0.010	0.020	0.051	0.103	0.211	4.605	5.992	7.378	9.210	10.597
3	0.072	0.115	0.216	0.352	0.584	6.251	7.815	9.348	11.345	12.838
4	0.207	0.297	0.484	0.711	1.064	7.779	9.488	11.143	13.277	14.860
5	0.412	0.554	0.831	1.146	1.610	9.236	11.071	12.833	15.086	16.750
6	0.676	0.872	1.237	1.635	2.204	10.645	12.592	14.449	16.812	18.548
7	0.989	1.239	1.690	2.167	2.833	12.017	14.067	16.013	18.475	20.278
8	1.344	1.647	2.180	2.733	3.490	13.362	15.507	17.535	20.090	21.955
9	1.735	2.088	2.700	3.325	4.168	14.684	16.919	19.023	21.666	23.589
10	2.156	2.558	3.247	3.940	4.865	15.987	18.307	20.483	23.209	25.188
11	2.603	3.054	3.816	4.575	5.578	17.275	19.675	21.920	24.725	26.757
12	3.074	3.571	4.404	5.226	6.304	18.549	21.026	23.337	26.217	28.300
13	3.565	4.107	5.009	5.892	7.042	19.812	22.362	24.736	27.688	29.820
14	4.075	4.660	5.629	6.571	7.790	21.064	23.685	26.119	29.141	31.319
15	4.601	5.229	6.262	7.261	8.547	22.307	24.996	27.488	30.578	32.801
16	5.142	5.812	6.908	7.962	9.312	23.542	26.296	28.845	32.000	34.267
17	5.697	6.408	7.564	8.672	10.085	24.769	27.587	30.191	33.409	35.719
18	6.265	7.015	8.231	9.391	10.865	25.989	28.869	31.526	34.805	37.157
19	6.844	7.633	8.907	10.117	11.651	27.204	30.144	32.852	36.191	38.582
20	7.434	8.260	9.591	10.851	12.443	28.412	31.410	34.170	37.566	39.997
21	8.034	8.897	10.283	11.591	13.240	29.615	32.671	35.479	38.932	41.401
22	8.643	9.543	10.982	12.338	14.042	30.813	33.924	36.781	40.289	42.796
23	9.260	10.196	11.689	13.091	14.848	32.007	35.173	38.076	41.638	44.181
24	9.886	10.856	12.401	13.848	15.659	33.196	36.415	39.364	42.980	45.559
25	10.520	11.524	13.120	14.611	16.473	34.382	37.653	40.647	44.314	46.928
26	11.160	12.198	13.844	15.379	17.292	35.563	38.885	41.923	45.642	48.290
27	11.808	12.879	14.573	16.151	18.114	36.741	40.113	43.195	46.963	49.645
28	12.461	13.565	15.308	16.928	18.939	37.916	41.337	44.461	48.278	50.993
29	13.121	14.257	16.047	17.708	19.768	39.088	42.557	45.722	49.588	52.336
30	13.787	14.954	16.791	18.493	20.599	40.256	43.773	46.979	50.892	53.672
40	20.707	22.164	24.433	26.509	29.051	51.805	55.759	59.342	63.691	66.766
50	27.991	29.707	32.357	34.764	37.689	63.167	67.505	71.420	76.154	79.490
60	35.535	37.485	40.482	43.188	46.459	74.397	79.082	83.298	88.379	91.952
70	43.275	45.442	48.758	51.739	55.329	85.527	90.531	95.023	100.425	104.215
80	51.172	53.540	57.153	60.392	64.278	96.578	101.880	106.629	112.329	116.321
90	59.196	61.754	65.647	69.126	73.291	107.565	113.145	118.136	124.116	128.299
100	67.328	70.065	74.222	77.930	82.358	118.498	124.342	129.561	135.807	140.170
125	88.029	91.180	95.946	100.178	105.213	145.643	152.094	157.839	164.694	169.471
150	109.142	112.668	117.985	122.692	128.275	172.581	179.581	185.800	193.208	198.360
200	152.241	156.432	162.728	168.279	174.835	226.021	233.994	241.058	249.445	255.264
1000	888.564	898.912	914.257	927.594	943.133	1057.724	1074.679	1089.531	1106.969	1118.948

NOTE: 3.927(-5) = 3.927×10^{-5} = 0.00003927

Table D $\mathcal{F}$–Distribution $\alpha = 0.05$.

Table entry is $F^* = \mathcal{F}_{\nu_1,\nu_2}(\alpha)$ where $P[\mathcal{F} > F^*] = \alpha$ (Tail area to the right)

	ν_1 = numerator degrees of freedom									
ν_2	1	2	3	4	5	6	7	8	9	10
1	161.45	199.50	215.71	224.58	230.16	233.99	236.77	238.88	240.54	241.88
2	18.51	19.00	19.16	19.25	19.30	19.33	19.35	19.37	19.38	19.40
3	10.13	9.55	9.28	9.12	9.01	8.94	8.89	8.85	8.81	8.79
4	7.71	6.94	6.59	6.39	6.26	6.16	6.09	6.04	6.00	5.96
5	6.61	5.79	5.41	5.19	5.05	4.95	4.88	4.82	4.77	4.74
6	5.99	5.14	4.76	4.53	4.39	4.28	4.21	4.15	4.10	4.06
7	5.59	4.74	4.35	4.12	3.97	3.87	3.79	3.73	3.68	3.64
8	5.32	4.46	4.07	3.84	3.69	3.58	3.50	3.44	3.39	3.35
9	5.12	4.26	3.86	3.63	3.48	3.37	3.29	3.23	3.18	3.14
10	4.96	4.10	3.71	3.48	3.33	3.22	3.14	3.07	3.02	2.98
11	4.84	3.98	3.59	3.36	3.20	3.09	3.01	2.95	2.90	2.85
12	4.75	3.89	3.49	3.26	3.11	3.00	2.91	2.85	2.80	2.75
13	4.67	3.81	3.41	3.18	3.03	2.92	2.83	2.77	2.71	2.67
14	4.60	3.74	3.34	3.11	2.96	2.85	2.76	2.70	2.65	2.60
15	4.54	3.68	3.29	3.06	2.90	2.79	2.71	2.64	2.59	2.54
16	4.49	3.63	3.24	3.01	2.85	2.74	2.66	2.59	2.54	2.49
17	4.45	3.59	3.20	2.96	2.81	2.70	2.61	2.55	2.49	2.45
18	4.41	3.55	3.16	2.93	2.77	2.66	2.58	2.51	2.46	2.41
19	4.38	3.52	3.13	2.90	2.74	2.63	2.54	2.48	2.42	2.38
20	4.35	3.49	3.10	2.87	2.71	2.60	2.51	2.45	2.39	2.35
21	4.32	3.47	3.07	2.84	2.68	2.57	2.49	2.42	2.37	2.32
22	4.30	3.44	3.05	2.82	2.66	2.55	2.46	2.40	2.34	2.30
23	4.28	3.42	3.03	2.80	2.64	2.53	2.44	2.37	2.32	2.27
24	4.26	3.40	3.01	2.78	2.62	2.51	2.42	2.36	2.30	2.25
25	4.24	3.39	2.99	2.76	2.60	2.49	2.40	2.34	2.28	2.24
30	4.17	3.32	2.92	2.69	2.53	2.42	2.33	2.27	2.21	2.16
35	4.12	3.27	2.87	2.64	2.49	2.37	2.29	2.22	2.16	2.11
40	4.08	3.23	2.84	2.61	2.45	2.34	2.25	2.18	2.12	2.08
50	4.03	3.18	2.79	2.56	2.40	2.29	2.20	2.13	2.07	2.03
60	4.00	3.15	2.76	2.53	2.37	2.25	2.17	2.10	2.04	1.99
70	3.98	3.13	2.74	2.50	2.35	2.23	2.14	2.07	2.02	1.97
80	3.96	3.11	2.72	2.49	2.33	2.21	2.13	2.06	2.00	1.95
100	3.94	3.09	2.70	2.46	2.31	2.19	2.10	2.03	1.97	1.93

Table E Studentized Range: $q_{\alpha_E,k,\nu}$ for $\alpha_E = 0.05$

	k = number of means								
ν	2	3	4	5	6	7	8	9	10
1	18.10	26.70	32.80	37.20	40.50	43.10	45.40	47.30	49.10
2	6.09	8.28	9.80	10.89	11.73	12.43	13.03	13.54	13.99
3	4.50	5.88	6.83	7.51	8.04	8.47	8.85	9.18	9.46
4	3.93	5.00	5.76	6.31	6.73	7.06	7.35	7.60	7.83
5	3.64	4.60	5.22	5.67	6.03	6.33	6.58	6.80	6.99
6	3.46	4.34	4.90	5.31	5.63	5.89	6.12	6.32	6.49
7	3.34	4.16	4.68	5.06	5.35	5.59	5.80	5.99	6.15
8	3.26	4.04	4.53	4.89	5.17	5.40	5.60	5.77	5.92
9	3.20	3.95	4.42	4.76	5.02	5.24	5.43	5.60	5.74
10	3.15	3.88	4.33	4.66	4.91	5.12	5.30	5.46	5.60
11	3.11	3.82	4.26	4.57	4.82	5.03	5.20	5.35	5.49
12	3.08	3.77	4.20	4.51	4.75	4.95	5.12	5.27	5.40
13	3.06	3.73	4.15	4.45	4.69	4.88	5.05	5.19	5.32
14	3.03	3.70	4.11	4.41	4.64	4.83	4.99	5.13	5.25
15	3.01	3.67	4.08	4.37	4.60	4.78	4.94	5.08	5.20
20	2.95	3.58	3.96	4.24	4.45	4.62	4.77	4.90	5.01
40	2.86	3.44	3.79	4.04	4.23	4.39	4.52	4.63	4.74
60	2.83	3.40	3.74	3.98	4.16	4.31	4.44	4.55	4.65
120	2.80	3.36	3.69	3.92	4.10	4.24	4.36	4.47	4.56

References

[1] Bogner, M. (2021). Applet: https://homepage.divms.uiowa.edu/~mbognar/applets/ for calculating probabilities from a variety of distributions. Department of Statistics and Actuarial Science, University of Iowa.

[2] Brochu, M., Poehlman, E., and Ades, P. (2000). Obesity, body fat distribution, and coronary artery disease. *Journal of Cardiopulmonary Rehabilitation*, **20**(2), 96–108.

[3] Fisher, R. (1925). *Statistical Methods for Research Workers.* Oliver and Boyd.

[4] Kelleher, J., MacNamee, B., and D'Arcy, A. (2020). *Fundamentals of machine learning for predictive data analytics: algorithms, worked examples, and case studies.* 2nd ed., MIT Press.

[5] McDougall, A. and Ahmed, S. (2017). Poster presented by MS STAT Thesis graduate Sahar Ahmed at the 5th Annual Canadian Statistics Student Conference. Poster and Workshop. Winnipeg, Manitoba.

[6] Moore, D., Notz, W., and Fligner, M. (2018). *The Basic Practice of Statistics.* 8th ed., W.H. Freeman.

[7] Myers, R. (2019). *Data Management and Statistical Analysis Techniques.* Scientific e-Resources.

[8] Orwell, G. (1945). *Animal Farm.* Secker & Warburg.

[9] Ott, R. and Longnecker, M. (2001). *An Introduction to Statistical Methods and Data Analysis.* 5th ed., Duxbury.

[10] Prezant, R., Chapman, E., and McDougall, A. (2006). In utero predator-induced responses in the viviparid snail bellamya. *Canadian Journal of Zoology*, **84**, 1–9.

[11] R Core Team (2021). R: A language and environment for statistical computing. https://www.R-project.org.

[12] Radelet, M. (1981). Racial characteristics and the imposition of the death penalty. *American Sociology Review*, **46**, 918–927.

[13] Ross, S. (2020). *A First Course in Probability.* 10th ed., Pearson.

[14] SAS Institute Inc. (2019). *JMP, Version 15.* SAS Institute Inc., Cary, NC.

[15] Thomas, D., Bredlau, C., Bosy-Westphal, A., Mueller, M., Shen, W., Gallagher, D., Maeda, Y., McDougall, A., Peterson, C., Ravussin, E., and Heymsfield, S. (2013). Relationships between body roundness with body fat and visceral adipose tissue emerging from a new geometrical model. *Obesity*, **21**(11), 2264–2271. https://onlinelibrary.wiley.com/doi/10.1002/oby.20408.

Index

Credits

Fig. 1.1: Copyright © by Microsoft.

Fig. 1.3: Copyright © by IBM Corporation.

Fig. 1.4: Adapted from New Jersey Department of Health, "Covid-19 Deaths in NJ: February to June, 2020," https://www.nj.gov/health/cd/documents/topics/NCOV/COVID_Confirmed_Case_Summary.pdf, State of New Jersey, 2020.

Fig. 1.5: Centers for Disease Control and Prevention, https://covid.cdc.gov/covid-data-tracker/#datatracker-home, 2022.

Table 1.2: Data source: R. Lyman Ott and Michael Longnecker, *An Introduction to Statistical Methods and Data Analysis*, p. 971. Copyright © 2001 by Cengage Learning, Inc.

Table 2.1: Data source: R. Lyman Ott and Michael Longnecker, *An Introduction to Statistical Methods and Data Analysis*, p. 971. Copyright © 2001 by Cengage Learning, Inc.

Fig. 3.12: R. Lyman Ott and Michael Longnecker, *An Introduction to Statistical Methods and Data Analysis*, p. 983. Copyright © 2001 by Cengage Learning, Inc.

Fig. 3.13: R. Clifford Blair and Richard A. Taylor, *Biostatistics for the Health Sciences*, p. 472. Copyright © 2008 by Cengage Learning, Inc.

Fig. 3.14: Source: https://www.census.gov/construction/nrc/historical_data/index.html.

Fig. 4.5: Copyright © by Matt Bognar.

Fig. 5.1: R. Lyman Ott and Michael Longnecker, *An Introduction to Statistical Methods and Data Analysis*, p. 614. Copyright © 2001 by Cengage Learning, Inc.

Fig. 9.1: Copyright © 2011 Depositphotos/leopolis.

Fig. 9.2a: Copyright © 2014 Depositphotos/artisticco.

Fig. 9.2b: Copyright © by The R Foundation.

Fig. 10.7: R. Lyman Ott and Michael Longnecker, *An Introduction to Statistical Methods and Data Analysis*, p. 153. Copyright © 2001 by Cengage Learning, Inc.

Fig. 10.8: R. Lyman Ott and Michael Longnecker, *An Introduction to Statistical Methods and Data Analysis*, p. 460. Copyright © 2001 by Cengage Learning, Inc.